*The Delaware Valley in the Early Republic*

CREATING THE NORTH AMERICAN LANDSCAPE

*Gregory Conniff, Edward K. Muller, and David Schuyler*
CONSULTING EDITORS

*George F. Thompson*
SERIES FOUNDER AND DIRECTOR

*Published in cooperation with the Center for American Places*
*Santa Fe, New Mexico, and Staunton, Virginia*

# THE *Delaware Valley*

## IN THE EARLY REPUBLIC

ARCHITECTURE,

LANDSCAPE, AND

REGIONAL IDENTITY

*Gabrielle M. Lanier*

THE JOHNS HOPKINS UNIVERSITY PRESS
*Baltimore and London*

*This book has been brought to publication with the generous assistance of Furthermore, a program of the J. M. Kaplan Fund, and the Mercantile-Safe Deposit and Trust Company.*

Photographs and diagrams not otherwise credited are by the author.

The Johns Hopkins University Press
2715 North Charles Street
Baltimore, Maryland 21218-4363
www.press.jhu.edu

*Library of Congress Cataloging-in-Publication Data*

Lanier, Gabrielle M.
The Delaware Valley in the early republic : architecture, landscape, and
regional identity / Gabrielle M. Lanier.
p. cm. — (Creating the North American landscape)
Includes bibliographical references and index.
ISBN 0-8018-7966-3 (alk. paper)
1. Delaware River Valley (N.Y.–Del. and N.J.) —History, Local. 2. Delaware
River Valley (N.Y.–Del. and N.J.) —Social conditions. 3. Delaware River Valley
(N.Y.–Del. and N.J.) —Ethnic relations. 4. Regionalism—Delaware River Valley
(N.Y.–Del. and N.J.) —History. 5. Ethnicity—Delaware River Valley (N.Y.–Del.
and N.J.) —History. 6. Human geography—Delaware River Valley (N.Y.–Del.
and N.J.) 7. Historic sites—Delaware River Valley (N.Y.–Del. and N.J.) 8. Landscape—Social aspects—Delaware River Valley (N.Y.–Del. and N.J.) —History.
9. Architecture—Social aspects—Delaware River Valley (N.Y.–Del. and
N.J.) —History. 10. National characteristics, American. I. Title. II. Series.
F157.D4L36 2004 2005
974.9—dc22 2004005666

A catalog record for this book is available from the British Library.

# CONTENTS

# ACKNOWLEDGMENTS

Many people have helped me throughout the course of this study. My first expression of thanks goes to my colleagues in the Department of History at James Madison University, especially Michael Galgano, David Dillard, David Owusu-Ansah, Jack Butt, Darryl Nash, Chris Arndt, Skip Hyser, and Steve Guerrier. Without their ongoing support and wonderful sense of humor, I would never have been able to complete this book. I am also grateful to Richard Whitman, Marilou Johnson, and the Faculty Assistance Committee of the James Madison University's College of Arts and Letters for awarding me a Summer Faculty Research Grant to support additional research for this project in its final stages. A generous dissertation fellowship supported by the McNeil Center for Early American Studies at the University of Pennsylvania allowed me to undertake the research and writing of several chapters when this book was still a dissertation; the center also provided a wonderfully stimulating intellectual environment in which to test my ideas. Among my colleagues there, Billy G. Smith deserves a singular note of thanks for his unfailing advice and encouragement when it mattered the most. I am grateful to Richard Dunn, director of the center, as well as to my fellow fellows Cynthia Van Zandt, Susan Stabile, George Boudreau and the wonderful community of scholars at the McNeil Center, which typically included Roger Abrahams, Roseanne Adderley, John Bezis-Selfa, Monique Bourque, Richard Chew, Jacob Cogan, Konstantin Dierks, Sally Griffith, Peter Hoffer, Daniel Kilbride, Susan Klepp, Roderick McDonald, John Murrin, Simon P. Newman, Leslie Patrick, Daniel Richter, Liam O. Riordan, Robert Blair St. George, Camilla Townsend, Stephanie Wolf, Sergei Zhuk, and Michael Zuckerman.

I am grateful to Gary Stanton for providing advice in the earliest stages of this study. Nancy Van Dolsen once again came to the rescue by providing an astute critical reading of the manuscript as it neared its final stages. Several individuals offered comments on earlier drafts, presentations, or conference papers based on this research. I am particularly indebted to Dell Upton, Joseph S. Wood, Liam Riordan, Garry Wheeler Stone, Carter Hudgins, the members of the Lancaster County Historical Society, and participants in Mc-Neil Center seminars at the University of Pennsylvania and Winterthur. Thanks also go to Pat Keller, Neva Specht, Cindy Falk, Tom Ryan, Ann Kirschner, Anna Andrzejewski, Graham Hodges, Camille Wells, Warren Hofstra, David Schuyler, and Jeanne Halgren Kilde for comments on presentations of early portions of several chapters. The perceptive comments of an anonymous reader for the Johns Hopkins University Press prodded me to push some of my ideas and produce a book that is, I hope, much improved over my original manuscript.

Members of the History Department at the University of Delaware were generous with their time and support when I was a graduate student there. I thank Patricia Orendorf, Marie Perrone, Janine Hobday, Christine Heyrman, Carol Hoffecker, David Pong, and Howard Johnson. Pat Elliott and Marguerite Connolly of the Winterthur Museum were always ready to answer my questions. My students at James Madison University, Millersville University, Mary Washington College, and Rutgers University may never realize how much they helped me to refine my thinking; they continue to inspire me in all kinds of ways. For discussions that often helped me to push or recast my ideas, I am especially grateful to Jill Kress, Susan Dawson, Stacey Schneider, Timothy Hack, Greg Kellerman, Shaun Mooney, Brian Dempsey, Peggy Dillard, Daniel Hayes, Jeanne Barnes, Ginny Harlow, Andrea Sarate, Mike Timonin, Jeff Shimabuku, Sarah Barrash Wilson, and Jamie Ferguson. At the University of Delaware's Center for Historic Architecture and Design, Rebecca Sheppard, Mark Parker Miller, Deidre McCarthy, and Jeroen van den Hurk offered good advice, kept me laughing, and were always there when I needed them. I have always especially valued David Ames's astute commentary and unique perspective.

For help with information on the various geographic subregions in this study, I thank Clarke Hess, Jim Waddington, Millie Dennis, Janet Sheridan, Dawn Melson, Michael Chiarappa, and David Schneider. Alice Guerrant and Gary Sachau of the Delaware State Historic Preservation Office, James Turk, David Young, and Barbara Duffy of the Salem County (New Jersey) Historical Society, and the staffs of the Lancaster County (Pennsylvania) Historical Society, the Chester County (Pennsylvania) Historical Society, the Chester

County Archives, the Maryland Historical Society, and the Delaware State Archives were all generous with their time and knowledge. The Interlibrary Loan department of James Madison University's Carrier Library cheerfully accommodated my many requests.

George F. Thompson and Randall B. Jones of the Center for American Places have been supportive at every step of the way. My teachers and many other friends and colleagues have helped to shape my ideas in more ways than they know. I am indebted to David Allmendinger, Kenneth Ames, Richard Bushman, James C. Curtis, J. Ritchie Garrison, David G. Orr, Kevin Sweeney, and Anne M. Whitmore. I wish that Alice Kent Schooler, who inspired my first research on the Chester County architectural landscape some time ago, had lived to see this project completed. Friends and colleagues in the Vernacular Architecture Forum have taught me—and continue to teach me—a great deal in the field. Particular thanks are due Carl Lounsbury, Nat Alcock, Marcia Miller, Peter Kurtze, Catherine Bishir, Phil Pendleton, Paul Touart, Willie Graham, Edward Chappell, Myron Stachiw, Michael Steinitz, Pam Simpson, John Larson, and, especially, Orlando Ridout V. I owe my most profound intellectual debt to Bernard L. Herman, who advised me during the dissertation stage, served as my most trusted field partner, and helped me to find my own voice.

My debts to my own family run enormously deep. I wish my father and Joshua Swain had lived long enough to see this book. Marjorie Milan, Albert and Lynne Perry, Robert, Virginia, Erica, Sarah, and Zachary Blickens, Erin and Steve Lutz, Amanda Swain, and Doug Brennan remained supportive to the end and gradually learned to stop asking when the book would be "done." My debts to my "other family," Bernie, Becky, and Lania Herman, run almost as deep, and I remain grateful for their perpetual friendship. Finally, I am grateful to the many property owners who opened their doors to me and gave so freely of their knowledge and good will. They always made fieldwork an enormous pleasure.

Writing projects often end very differently from the way they began. This one acquired a life of its own midway through the process, and ended up diverging significantly from the original idea that formed its catalyst. It began as a relatively straightforward attempt to examine the regional identity of the lower Delaware Valley in the early national period through several architectural and landscape case studies. It ended up as a story that proceeded not from within a single discipline but from the intersections of American social history, cultural geography, and material culture studies, unfolding gradually, not only in the realm of the material landscape, but also in the unevenly charted territories of environmental perception, cultural mappings, and public memory.

Historians are accustomed to dealing with the biases inherent in their evidence. Letters, tax lists, court records, inventories, and all kinds of other documents survive because they have been deemed important, because they constitute part of the public record of significant events, because they were thought worthy of preservation, or they have survived by accident. Buildings are no exception: those early buildings that survive often do so only through the efforts of preservation-minded individuals, because they belonged to someone "important," because they were somehow connected with significant local, regional, or national events, or, occasionally, by accident. Yet any time one uses the built environment as evidence, one is made aware of more subtle factors at work, factors that transcend the mere survival and preservation of buildings and material objects.

Dwellings are more than just shelters. To observers, buildings and the landscapes that surround them can suggest social relationships or cultural affinities. Because a house with its landscape usually constitutes one of the most visible material records of an individual's life, it comes to signify something about the way its inhabitants view their world, or what they hope to become. Houses also mark their inhabitants as members of, or outsiders to, a particular kind of community. And buildings, because they are so visible and monumental, also serve as physical points of reference in the landscape as well as touchstones for public memory about place. In the aggregate, they form visual communities that express approaches to or ways of viewing the landscape. But, as most architectural surveyors know only too well, while buildings may last, context—the interstices or "connective tissue" between buildings—is one of the first features to disappear. Often, then, we are left with a shell—a significant portion of the material record, but a portion nonetheless. And just as often, we tend to fill in the interstices with our own perceptions, just as period observers did.

Missing physical context, of course, can be reconstituted to an extent. From historical documents and material evidence, we can frequently reconstruct fuller representations of historic built environments that encompass ephemeral features such as outbuildings, fences, and field patterns. Yet context is perceptual as well as physical, and it involves values, reactions, and interpretations as well as buildings and landscapes—in short, our responses to the material world affect the way we interpret it. Although, from a historian's point of view, this perceptual context is one of the most difficult features to get at, it is also one of the most helpful tools for understanding and imagining historic landscapes, for, as Dell Upton has argued in his study of early Philadelphia, it is the connection *between* person and artifact rather than the artifact itself that demands the closest scrutiny when studying historic urban landscapes.[1]

Period observers, too, colored the artifactual environment with their perceptions and responses to it. Yet perceptions change, and artifacts, even those as large as landscapes, can gain and lose meaning through time. And although the meanings that were once invested in past landscapes may differ significantly from our present-day reactions to the same environment, past perceptions and reactions can continue to shape contemporary interpretations. This reflexive aspect of the cultural landscape—this ongoing dialogue between historical and contemporary landscape perceptions—not only constitutes a critical component of past and present mental maps, but also stands at the heart of the process of landscape formation. In turn, this mental mapping process is critical to the process of forming what we think of as "regional identity."

Pivotal to this study is the concept of a historical regional identity in the

Delaware Valley as it was expressed in and informed by the material world. "Identity," at least as a dictionary would define it, usually connotes a kind of unity, consistency, or sameness of essential character; a "regional identity" similarly implies some level of cohesion in the way that distinctive characteristics are expressed or regarded within a particular geographic area. Yet "identity," like "culture," "ethnicity," and "region," is a slippery term with layered meanings and indeterminate borders. The study of something as elusive as a past regional and social identity depends upon both intentional and nonintentional characterizations of self and environment by people in the past—representations that typically issue from written and visual documents as well as the material record.

Another central organizing concept in this study is the notion of cultural landscape, which is defined here, in Peirce Lewis's terms, as the sum total of the changes people have wrought upon their environment, or "a kind of cultural autobiography that humans have carved and continue to carve into the surface of the earth."[2] While this study explores the identities expressed in three Delaware Valley cultural landscapes, it also coalesces around three primary ideas, all of which relate to the ways groups of people invest meaning in their shared physical spaces. The first of these ideas is the notion that meanings invested in objects sometimes change as a result of cultural circumstance. In other words, when cultures are under pressure to change in some way, they may endow objects with deeper meaning and begin to use material culture in new, and sometimes different, ways. This idea becomes especially relevant during the period of widespread change following the American Revolution. The second idea centers on the ways in which cultural boundaries are established, maintained, and negotiated in material form—or, put a little differently, the ways in which expressed identity or resistance to change might have been articulated or consolidated. The third revolves around the notion of landscape prosopography, "thick landscape descriptions," or what the writer William Least Heat-Moon has termed "deep maps." A deep map is what Heat-Moon calls "a topographic map of words"—a thick description of place that breathes life into the two-dimensionality of the map and that transcends the purely cartographic or the simply descriptive to become a fusion of history, narrative, memory, and imagination.[3] This essay aims to interpret and contextualize three cultural landscapes and to build a deep historical map of these parts of the lower Delaware Valley in the early national period.

The Delaware Valley offers an ideal arena for testing this notion of a cohesive regional identity since the cultural diversity of this area was historically extreme. Yet despite this historical diversity, historians, geographers, and folklorists have often portrayed the area as a definable region that is somehow supposed to fill the conceptual space between the well-delineated historical and geographical characterizations of New England and the Chesapeake.[4] Still, as anyone who has undertaken architectural fieldwork in the area can attest, the built environment tells a different story, for the area's architectural landscape in the early days of the Republic was far less regionally cohesive than many of these characterizations suggest. The Delaware Valley was instead a rich cultural crossroads that still exhibits an astonishing diversity in the built landscape, not one region at all but a region of regions. Even late in the twentieth century, geographer Wilbur Zelinsky noted that the Middle Atlantic was the "weakest and loneliest" of all of the definable vernacular regions in the United States.[5] In the early national period, subregions in the area ranged from relatively stable and homogeneous locales to shifting, speculative urban landscapes, from areas with long-established traditions of cultural diversity to townships whose cultural traits were established primarily by settlers transplanted to the region from New England.

While late-twentieth-century suburban expansion and development have altered some of these early architectural landscapes, many well-preserved fragments survive in surprising proximity to heavily urbanized areas. In fact, one of the pleasures of undertaking fieldwork in the Delaware Valley has always been the nearness of these diverse early landscapes: from northwestern Delaware, one can drive for an hour or two in any direction and encounter a completely different architectural and landscape subregion. Heading to the northwest, one crosses into the Lancaster Plain, with its fertile open countryside, distinctive bank barns, and German-influenced architectural traditions. Moving more directly north into the Piedmont, toward the borders of western and northern Chester County, Pennsylvania, one encounters instead the collision of Anglo, Scots-Irish, and some Germanic architectural traditions. Moving further northeastward into the heart of Chester County, one crosses over into the gently rolling landscape of the Welsh Tract, with its strong Welsh and English imprints. Even further to the east, in the heavily urbanized landscapes of Philadelphia's Northern Liberties, one can uncover evidence of the speculative "edge" landscapes that characterized the city's rapidly expanding semirural fringes in the early national period. Directly eastward, just across the Delaware River in the marshy lowlands of New Jersey's inner coastal plain, one is among English Quaker landscapes and a distinctive subregional building tradition issuing from an early in-migration of New England settlers. Further to the northeast, in the Dutch-settled regions of New Jersey's outer coastal plain, one can see portions of the Dutch architectural imprint. Changing direction again and heading southward down the Delmarva Peninsula to lower Delaware and Maryland's Eastern Shore, one is confronted by an entirely different landscape. Although the flat and marshy countryside here resembles that of New Jersey's inner coastal plain, the architectural idiom has changed once again. Here, although little of the early architectural landscape survives, one encounters occasional physical remnants and extensive documentary evidence of the Chesapeake building traditions that once prevailed.

This diversity in the built landscape mirrors the extreme cultural diversity for which the area has always been known: the Delaware Valley constituted a mosaic rather than a monolith. While thorough analysis of any number of the area's distinctive subregions would densely illustrate this diversity, it would also add unnecessary layers to a picture that is already complex enough to belabor the most obvious point. Consequently, the following study focuses on three primary landscape subregions that address this basic diversity and raise further issues central to the consideration of regional identity. These particular subregions have been selected to provide the broadest possible cross-section of the area's cultural diversity and because they express some of the various factors that shape cultural identity, such as strong ethnic, class, or religious and sectarian imprints, racial diversity, political orientation, economic organization, or distance to or from major markets. But while each of these three areas forms the geographic focus of a chapter, the boundaries of cultural subregions do not necessarily follow the arbitrary boundaries of townships, hundreds, or political jurisdictions. Since the borders of such subregions are typically blurred and indistinct, evidence from surrounding geographically and culturally similar areas has also been utilized to develop the discussion.

"Region" and "identity" have been construed in many ways. Similarly, the factors that contribute to form a regional identity can range from environment and economy to politics and belief. Chapter 1 explores the idea of the Delaware Valley as a region, examines the notions of "region" and "regional identity," and explores how these concepts have been used to inform the study of the Delaware Valley and Mid-Atlantic region during the early national period.

Since ethnic composition is often conflated with regional identity, Chapter 2 focuses primarily on Warwick, a strongly Germanic township in Lancaster County, Pennsylvania.[6] Geographers, historians, and folklorists have long maintained that, along with New England and the lower Chesapeake, southeastern Pennsylvania constitutes one of the country's three major cultural hearths from which cultural traits eventually diffused outward. Fred Kniffen has argued that, with its early and dense settlement of Germans, English, and Scots-Irish, the southeastern Pennsylvania hearth had the most widespread influence of the three.[7] Warwick Township, situated just north of the major inland market town of Lancaster, stands at the very center of this critical area. Overwhelmingly populated by settlers of Germanic origin during the colonial and early national periods, Warwick also includes the Moravian borough of Lititz near its geographic center. The township's historically strong Germanic concentration, especially visible in its surviving architecture, permits the study of a subregion that might be described as ethnically homogeneous. Still, because not all parts of southeastern Pennsylvania were as homogeneous as Warwick, several more heterogeneous subregions have also been considered here for the sake of comparison. Nearby Hempfield and Conestoga Townships, both in Lancaster County and both heavily influenced by Germanic settlement, nonetheless exhibited greater ethnic diversity than Warwick by the end of the eighteenth century. Even more diverse was Sadsbury Township, east of Lancaster County and on the western border of predominantly Quaker Chester County. Advantageously located on the main market route between Philadelphia and Lancaster, Sadsbury was inhabited primarily by English and Scots-Irish but was close enough to the more strongly Germanic regions of Lancaster County to be affected by that area's culture and building practices as well. It therefore occupied something of an ethnic and economic cross-roads.

North West Fork Hundred, far to the south of Warwick in Sussex County, Delaware, forms the focus of Chapter 3. The material and social landscapes of this part of Delaware differed substantially from other parts of the lower Delaware Valley in the early national period, for North West Fork is the southernmost landscape subregion in the study, and the one most closely tied to Chesapeake cultural and building traditions. North West Fork is at the midpoint of the Delmarva Peninsula, abutting Maryland's Eastern Shore. Once a large slaveholding region, North West Fork Hundred originally formed a part of neighboring Dorchester County, Maryland, until the state boundary dis-

pute was resolved in 1776. Even as late as the early 1800s, this politically contested area still occupied something of an economic and geographic margin,
but, just as significantly, it was also regarded by contemporaries as a racial, cultural, and conceptual border zone in both factual and fictional accounts. Although few examples of architecture from the earliest period of durable building survive in this region, documentation describing the built environment is
extensive and permits a detailed reconstitution of the early national landscape. Since the surrounding Sussex County economy and built environment
as well as adjacent parts of Maryland's Eastern Shore closely resembled North
West Fork in many respects, documentary evidence from these areas has also
been included.

Religious and sectarian imprints are also often implicated in the formation
of regional identities, especially in the case of the Delaware Valley. Because the
entire region has frequently been characterized by other historians as being
largely shaped and controlled by the Quaker influence, a relatively homogeneous, Quaker-settled landscape has been included through which to explore
this assumption. Mannington Township, on the east side of the Delaware River
in Salem County, New Jersey, constitutes the focus of Chapter 4. Mannington
represents an intensively settled, well-established and largely culturally homogeneous agricultural subregion dominated by English Quakers. Still, this
dispersed agricultural landscape was controlled by a powerful minority, a native-
born Quaker elite, whose substantially built brick houses, many of which survive today, soon became conspicuous statements of success. As with the other
subregions included here, evidence from several nearby culturally and environmentally comparable jurisdictions is introduced to augment the discussion. Nearby Greenwich Township, located in adjacent Cumberland County,
New Jersey, and also abutting the Delaware River, provides an important example of transplanted cultural practices, many of which are strikingly evident
in the area's architecture. Like Mannington, Greenwich was settled early, but
its early cultural landscape was the product, not so much of a longstanding
English Quaker imprint, but of a significant early in-migration southward
from Rhode Island and other parts of New England. Physical evidence of this
migration still survives in the anomalous southeastern New England framing
traditions visible in numerous dwellings dating to the earliest period of
durable building. Similarly, evidence from Lower Alloways Creek Township
and other nearby parts of Salem and Cumberland Counties has been used to
flesh out the examination of Mannington.

The final chapter places these three primary landscapes within the broader context of the Delaware Valley and the Mid-Atlantic and explores the interplay between localism and regional and national identities in the early national period. The timing of the emergence of American regional and national identities has been a matter of some debate. Some historians argue that the development of American regional identities was closely linked to the nationalizing impulse.[8] Recently, Edward Ayers and Peter Onuf have suggested that, while consciousness of difference and regional identity were direct outgrowths of the nationalizing impulse, actual local and regional differences always existed. The more important question to ask, they suggest, is *why* certain cultural distinctions came to matter more than others in the quickening construction of collective identities after the Revolution. In other words, they suggest, the choices that have been made and the *process* of forming such regional identities may reveal just as much as the identities themselves.[9]

The three cultural landscapes examined here enable us to consider, to varying degrees, that very process of identity formation. With its remarkable ethnic, sectarian, and cultural diversity in the early national period, the Delaware Valley—in many ways the new nation writ small—provides a unique laboratory for exploring some of these issues. These three landscapes define this "region" as a mosaic of localized places, or a true "region of regions." Equally significantly, though, because these landscapes effectively exemplify some of the underlying tensions between localism and regionalism, they also allow us to consider how local, regional, and national identity might intersect with several more current analytical categories that may have superseded regionalism as viable constructs for studying the past.[10] By considering how some of these analytical categories are embedded within these localized cultural landscapes, then, this essay not only complicates existing interpretations of the Delaware Valley and lower Mid-Atlantic that use regionalism as an organizing construct but also reevaluates this "weakest and loneliest" of American vernacular regions through multiple and sometimes overlapping lenses.

*The Delaware Valley in the Early Republic*

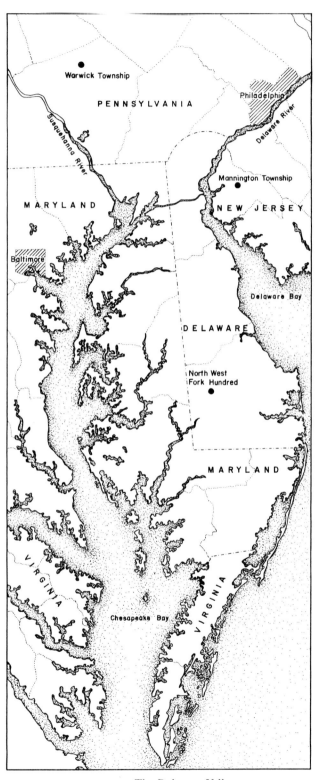

FIG. 1.1. The Delaware Valley

# CHAPTER 1

# "The Motley Middle"

WARWICK TOWNSHIP, LANCASTER COUNTY,
PENNSYLVANIA, 1815

Halting in front of Christian Becker's substantial two-story limestone house, the tax assessor began to record specific details about the farmer's property. The main dwelling backed against a slight hill, over-looking 194 acres of mostly cleared, rolling, and fertile Lancaster County land. The house had four rooms on the ground floor, but it had always been heated with stoves rather than fireplaces, and its interior was furnished with the same flat, vasiform stair balusters that ornamented several other Pennsylvania German houses in the area. In addition to the main dwelling house, the assessor counted five other dwellings, four barns, a tenant house, and two stables. Two of Becker's barns were par-ticularly impressive: his older wood-and-stone barn measured nearly 100 feet in length, and his brand new overshot bank barn resembled several other new barns in the area in terms of size as well as functional design— it was built entirely of stone with enough room inside to turn wagons around easily.

### GREENWICH TOWNSHIP, CUMBERLAND COUNTY, NEW JERSEY, CIRCA 1700

The joiner marked the ends of each of the house timbers with Roman numerals. Just as all the other joiners in southwestern New Jersey did, he cut square mortises into the corners of the sills where they were lapped, so that tenons at the bottoms of the corner posts could fit into these mortises. But when he began forming the joints for the tops of the corner posts, he fashioned a complex shouldered joint which was uncommon in the repertoire of other local joiners. He remembered seeing this joint on a number of house frames in Rhode Island several years earlier and had already used it to good effect on other houses in the Greenwich area.

### PHILADELPHIA, PENNSYLVANIA, 1803

James Vaux laid out on the table the plan showing all of the new building lots. He was offering several groups of lots for sale, on various streets in the rapidly developing Northern Liberties section of the city, all suitable for building the small, two-story houses that seemed to be sprouting everywhere in this part of Philadelphia. There were one hundred nineteen lots in all. Some of the choicest properties stood on Callowhill Street. These were ideally suited for artisans' shops and other businesses because most extended from the street in front to Pegg's Run, and a new market house was planned nearby. Within this select group of Callowhill lots stood a smaller cluster of six properties that fronted on a 30-foot-wide alley. Hoping to make as much profit as possible in this burgeoning market, Vaux offered to sell this smaller group of lots "separate or together as may suit the purchaser."

### SADSBURY TOWNSHIP, CHESTER COUNTY, PENNSYLVANIA, 1789

Samuel Wilson began laying the stone foundation walls of John Truman's new house. The house, located on the fertile valley bottom near Buck

Run Creek, would be two stories high, nearly square in plan, with three openings across the front, two end chimneys, and four rooms on the ground floor, each containing a fireplace. The cellar alone had taken Wilson and his son Joseph nearly four days to excavate. He had quarried and hauled tons of local quartzite and schist to the site to build the coursed rubble walling, chimney stacks, hearths, and window lintels. He had carted lime burnt in nearby limekilns to make the mortar, and then spent a day and a half working on the cellar windows. Wilson was looking forward to the day when Truman's house would be finished, for it would complement the group of stone buildings he had already constructed for this prosperous Quaker farmer. In addition to a stone mill and barn nearby, Wilson had built a small, arched stone springhouse for Truman five years earlier. Springhouses were common in Sadsbury because water sources were plentiful and dairying was especially important to the local economy. Still, the one that Wilson built for his client was a bit unusual because it was one of only two in the area that were roofed with sod.

NORTH WEST FORK HUNDRED, SUSSEX COUNTY,
DELAWARE, 1807

On a clear February day, William White and Leon Todd trudged across the flat, stubbled cornfield to Manlove Adams's one-story frame house. Adams had died without a will, and Todd and White were charged with recording all his worldly goods. Although Adams had been one of the wealthier individuals in the area and his 16-by-20-foot house was still in reasonably good repair, the other buildings on the property were in decidedly sorry condition. His property also included a log house with an earthen floor and a small milkhouse, both of which were dilapidated. Adams's household furnishings were equally modest: the inventory takers noticed several beds, a few tables, an old pine chest, a cupboard, some kitchen goods, and some textile equipment. By far, the most valuable items of property in Adams's estate were his carriage, his livestock, his stored corn in the crib, and, of course, Unice, Elijah, James, John, Draper, and Easter, his six slaves.

## MANNINGTON TOWNSHIP, SALEM COUNTY, NEW JERSEY, 1770

The small group of men gathered around the table were in agreement: they would stake out and mark where the drainage ditches would be cut and the sluices laid; then they would determine how to finance the labor and upkeep. This additional work was necessary to keep the low-lying tidal marshes bordering Mannington Creek drained and productive for all of them in the coming year. Everyone in the group owned property abutting the creek, and many of them pastured their cattle behind tide banks or cut salt hay on their reclaimed meadows. Most of them were wealthy landowners, many of English Quaker extraction. Ever since that day back in 1756 when a small handful of them had first joined forces at William Hall's imposing two-story brick house, this group had been meeting annually to choose their officers and to determine what work needed to be undertaken on the reclaimed meadows. It was good that these men had decided to band together; acting collectively, it was much easier for them to finance significant improvements and exert some real control over the productivity of the Salem County marshlands.[1]

These six very different early American cultural landscapes all fell within the compass of that region known now as the Delaware Valley or the lower Mid-Atlantic. All lie in close proximity to central Philadelphia, within a morning's drive today. Although urban sprawl and modern transportation networks have effectively muted many of the boundaries that once existed between these earlier landscapes, their distinctiveness in the early national period and the first few decades of the nineteenth century remained significant, not only in terms of population, economy, and ethnic and racial mix, but, perhaps just as importantly, in terms of the messages from the material world. The Delaware Valley or lower Mid-Atlantic—the broad geographic swath that includes Delaware, southwestern New Jersey, and the portion of southeastern Pennsylvania that lies east of the Susquehanna River and just north and west of Philadelphia—falls within that gray and troublesomely ill-defined midsection of the country that historians can neither characterize as "Puritan" New England or "Cavalier" South. Although this area has always been known for its cultural diversity, scholars have often persistently characterized the Delaware Valley monolithically as a broadly definable region or else simply allowed the far more well-delineated historical characterizations of New England and the Chesapeake either to overshadow it or to form the boundaries of another region by default. Yet, at least as reflected in the material record, during the early national period the Delaware Valley remained a collection of very distinctive and intensely local subregions, each with its own building traditions, populations, land use patterns, material culture, and accompanying landscape perceptions.

This material localism mirrored the area's exceptional cultural diversity, a point that was not lost on contemporary observers. Visitors repeatedly remarked on the area's cultural pluralism and how that was manifested in social and material ways. When Thomas Cooper described Philadelphia's bustling, multiethnic, and multilingual diversity in 1794, he remarked in particular on the "emigrants of every class that come to the port of Philadelphia." To ensure success in business, Cooper recommended, aspiring British merchants, tradesmen, and shopkeepers coming to America should also know both German and French. On shop signs in Philadelphia, he wrote, "every storekeeper has the name of his firm, and his trade, written in the German character and language, as well as the English." Similarly, when Isaac Weld described the mostly German inland market town of Lancaster in 1800, he remarked that, of a total of six churches, five were designated for different denominations: "Of the churches, there is one respectively for German Lutherans, German Calvinists, Moravians, English, Episcopalians, and Roman Catholics."[2] Despite this acknowledged diversity, monolithic characterizations of the area tend to diminish its significant cultural plurality and mask the nuances of what was, in fact, a very complex early American material and social landscape. The region's architectural, landscape, and cultural pluralism, which persisted well into the early decades of the nineteenth century, not only challenge the monolithic characterizations of the region that still prevail in much of the existing historical literature, but also suggest that localism rather than regionalism may be a far more potent organizing concept for approaching the study of American history and culture in the early national period.

The period between 1780 and 1830 represents an especially suitable time span in which to explore the expression of a regional identity in the Delaware Valley built landscape. While these years represent a formative time in the national history of of the early republic, at a more local level, they also represent the beginning of the second hundred years of Anglo-European settlement for much of the region. By the close of the eighteenth century, most of the area had been settled for a hundred years or more, and the dwellings of the mid-eighteenth century constituted aging housing stock in an increasingly changing landscape. The built environment consequently experienced significant recasting during this period, as new buildings were constructed and old ones were expanded or reworked. By the end of the century, too, the first waves of inhabitants had long since established their initial cultural imprint on the countryside through settlement patterns and land use practices as well as buildings, and new generations inherited and redefined these material expressions. As ethnically diverse populations expanded following successive waves of settlement throughout the eighteenth century, spatial intermingling between various cultures increased. In southeastern Pennsylvania, for example, diverse ethnic groups, including English, German, Scots-Irish, and Welsh inhabitants, and equally diverse religious groups, such as German Dunkers, Mennonites, Lutherans, Welsh and Scottish Anglicans, Reformed Huguenots, and English Quakers overlapped in many areas by midcentury.[3] The nationalizing changes wrought by the Revolution only augmented this regional intermingling. Also, while the Revolution had fractured numerous long-established political, ethnic, and religious connections, the period of change that followed encouraged a reevaluation of many of these distinctions. This transformative period, then, was one during which the nationalizing impulse encountered the physical and conceptual remnants of earlier cultural landscapes. As America entered a period of conscious nation-building during these years of the early republic, local and regional culture was increasingly— if fitfully—becoming "American" culture.

Regionalism has long been a compelling and persistent theme in American scholarship. Geographers, historians, linguists, folklorists, anthropologists, journalists, and other observers have reflected upon American regions, their boundaries, their cultures, and the impact they have had on our individual and collective identities. Despite the diversity of American experience, the notion of localized places of belonging—places that express our unity rather than our diversity—continues to sustain us.[4]

In America, regional self-consciousness has followed cyclical historical rhythms, with certain periods witnessing especially widespread interest in regional concerns.[5] During the 1920s and 1930s, especially, regional thought flowered and became a topic of concern to scholars and practitioners in many disciplines. Regional thought and ideology flourished then among historians, planners, artists, folklorists, engineers, architects, and social scientists, and the region as a construct assumed a new utility.[6] The heightened interest in regional differentiation during these decades has been attributed, in part, to the ongoing debate over the nature and timing of American nationalism and cultural pluralism and their relationship to the linked processes of urbanization and surging European immigration to the cities. Some scholars have also argued that the growing interest in regionalism stemmed from the equally important need for a sense of place during the stressful Depression years.[7]

This blossoming of interest in regionalism as a valid and resonant cultural construct developed in tandem with concern that regional differences were declining—that traditions were threatened, folk values were imperiled, cultures were dispersing, and an "older," mostly rural, America was rapidly disappearing. But "region," to many of these scholars, meant a way of life as well as a particular place. While a critique of modernization and its concomitant transformations was inherent in much of the regionalist movement of the 1930s, an important aspect of this trend was its implicit promise for the future as well as its link to the past. With its emphasis on local folkways and environments, the region was seen during this period as a way to mitigate the nationalizing, homogenizing effects of urbanization, industrialization, and modernism. The hope was that the idea of region would provide a blueprint for new kinds of more liveable, approachable cities and reinvigorated rural economies.[8] For many of these regionalists, then, "region" was both a sociological and an ideological construct, a state of mind as well as an agglomeration of states. And therein lies one of the difficulties inherent in any examination of regionalism and regional identities, for, as pervasive as the idea of region has been throughout American thought, it still remains a complex notion to define.

The word "region" usually carries with it a spatial connotation. Most discussions of region have at their root the idea of a particular geographic entity or physical place. A region in this sense is customarily defined and bounded by geographic features such as mountains or rivers. Yet regions can also be defined by political boundaries, economic constraints, cultural practices, or any combination of these factors. Howard Odum and Harry Estill Moore began their landmark 1938 study of American regionalism with an entire page that included more than forty varying definitions of the concept of region. These included, among others, "a spontaneous expression of physical and psychological differences," "an area . . . delineated on a basis of general homogeneity of land character and general homogeneity of occupance," "an area where nature acts in a roughly uniform manner," and "a territory of indefinite extent."[9] Nonetheless, regional boundaries defined by one set of criteria often diverge from those defined by other factors, and regions distinguished by cultural traits frequently spill beyond the crisp boundaries of political jurisdictions or topographical features.

The folklorist Henry Glassie made the distinction between cultural regions and geographic or political regions when he wrote "a cultural region is not an isolated pocket where funny old stories may be gathered in, nor is it an area marked off by arbitrary political boundaries or a noncultural clutch of topographical features." A cultural region, Glassie continued, is defined by cultural commonalties. Similarities in language, cultural practices, or material objects describe the core and delimit the boundaries of cultural regions.[10] Glassie used the spatial and temporal distributions of features, such as house plans, agricultural practices, and farmyard plans, to map six contiguous material folk culture regions in the eastern United States. Still, he acknowledged the difficulty of defining such cultural regions in precise geographic terms, for they have "fuzzy, syncretistic borders and any attempt to define them on a map is a process of constant compromise."[11]

Like Glassie, other scholars, including Hans Kurath and Fred Kniffen, have wrestled with the spatial definition of cultural regions. Kurath, whose conception of region is based on the notion of cultural phenomena diffusing outward from a cultural center, mapped the areal distribution of dialects in the eastern United States to reveal clearly differentiated regionalisms in speech. To Kurath, the diversity of speech on the Atlantic seaboard was the result of geographically restricted early settlements, and the gradual disappearance of regional dialects was due to the progression from local cultures to regional and, finally, to national cultures.[12] The assumptions here, which are widespread in many such discussions of regional distinctiveness, are that regional distinctions, while inherited from the past, would gradually but inevitably be adulterated through time and that such differences are often closely linked to differences in land and climate.[13] Similarly, Fred Kniffen plotted the spatial distribution of building types to reveal three primary cultural source areas on the Atlantic seaboard: New England, the Middle Atlantic, and the lower

Chesapeake. Kniffen's findings, like those of Kurath, are based on the notion of outward diffusion from cultural centers, or hearths. Kniffen's famous map of diffusion routes issuing outward from these cultural hearths suggests that, of the three main source areas, the Middle Atlantic area had the most far-reaching influence.[14] Still, despite the acknowledged importance of this area, many observers agree that its identity remains indistinct. While Glassie's assessment of American material folk culture regions concurred with the profound importance of Kniffen's Mid-Atlantic cultural hearth, Glassie also noted that this was initially the "least homogeneous" region of all. Similarly, the geographer Wilbur Zelinsky, writing about North American culture areas, has noted that a "muted vagueness about their collective personality" has always plagued the inhabitants of this Pennsylvania culture area.[15]

Also embedded in this concept of a cultural region, however imprecise its geographic boundaries may be, is the notion of tradition, which surfaces in a variety of ways. Tradition is expressed most simply in opposition to innovation or change. In their discussions of diffusion, Kniffen and Kurath focus on the persistence and diffusion of "traditional" cultural patterns from a parent culture across a particular geographic area. Traditional culture is typically diffused outward, away from some sort of a center or core or hearth, sometimes being mediated by other cultures or other influences such as modernization or technological change. Environment, too, plays a part in mediating culture. In fact, this notion of a "traditional" region often seems strongest in areas where the tie to the land is greatest and where any encroachment by "modern" life is least apparent.

Still, as some of these scholars have acknowledged, cultural regions remain difficult to define with certainty, not only because their geographic boundaries are sometimes blurred, but also because a "cultural region" can be viewed in several different ways. In his study of the regional consciousness embedded in the artifacts of the Connecticut River Valley, Robert Blair St. George points out that one reason the concept of regional culture itself remains an inexact tool for analysis is that its definitions are so imprecise. Confusion arises because theorists of regional culture have used at least five different criteria for defining regions, including natural features, metropolitan proximity, folkways, political or bureaucratic convenience, or broad subnational geographic relationships. Additional types of cultural regions also emerge in various disciplinary approaches to society and culture, including "the 'functional regions' of geographers, the 'mercantile regions' of economists, the 'administrative regions' of political scientists, and the 'aesthetic and literary regions' of art historians and literary critics."[16] Other scholars have further complicated the picture by defining cultural regions on the basis of prevailing attitudes toward the land, the way power has been exerted on and within a particular landscape, or even the prevalence of ideology and kinship networks.[17] The reciprocal relationship between culture and region also remains a critical factor. As Wilbur Zelinsky has commented, "a particular culture, or combination of subcultures, helps impart to an area much of its special character and behaviorial design," and conversely, "the character of a given place may be a strong formative influence in the genesis of a particular culture."[18]

However a regional culture may be delineated in geographic, political, or economic terms, its boundaries remain critical to its identity. In fact, the cultural identity of a region may be most intense at its boundaries rather than at its center. This becomes especially notable during periods of widespread and rapid change, such as the decades following the American Revolution. During such times of instability and change, cultural boundaries are often the locus of intense definitional activity. And, as several scholars have noted, the material expressions of a culture often become the most assertive at the point when that culture is experiencing the greatest pressure, "when communities have to drop their heaviest cultural anchors in order to resist the currents of transformation."[19] Yet actual distinctions and awareness of those distinctions are not the same. Perceptions of regional cultural boundaries can also contribute to defining a regional culture. Such perceptions often become important because they become embedded as stereotypes. Nonetheless, as the anthropologist Anthony Cohen has reminded us, all perceptions are not the same, and "the boundary may be perceived in rather different terms, not only by people on opposite sides of it, but also by people on the same side."[20] Still, it is the consciousness of local and regional distinctions, the awareness on the part of both insiders and outsiders of just where the cultural boundaries are, that contributes to defining a regional identity.[21]

Like cultural boundaries, "identity" may also be perceived differently depending on one's spatial or historical point of view. While "identity" implies a unity, consistency, or sameness of essential character and may be defined loosely as the way an individual or group is perceived or wishes to be perceived by others, the actual interpretation of an identity may vary. For culture, as Anthony Cohen argues, cannot be magnified and reified as a "monolithic determinant" of human behavior: "If culture did have that character, it would equip us with uniform rather than with identity. Culture is a matter less for documentation than for interpretation; it is more faithfully and sensitively depicted in metaphor than in museums. Its intellectual fascination lies in its extraordinary versatility, which is precisely what makes it such an eloquent representation of identity."[22]

How, then, does a "regional identity" come about? If a regional identity equates to a measure of consistency in the way a particular place is perceived, then what factors work to shape it, and when, in terms of historical development, does such a regional identity become possible? Factors such as land and climate, ethnicity and race, economy, politics and government, nationalism, culture, wealth, settlement history, and market orientation have long been recognized as contributors to regional and national identity; more recently, the roles of gender and religion have been explored.[23] But some have also argued that regions themselves are constantly changing, and their relationships to other regions are always changing, too. Attitudes toward place shift through time, and history, which imparts multiple layers of meaning to the environment, plays a critical role in this process. In other words, regions are continually being redefined, and this ongoing redefinition of actual regions also affects regional identities. Some scholars have maintained that American national identity emerged before regional identity, and that consciousness of difference was a direct reaction to the formation of the nation-state and a product of a newly formed common context. Consequently, in order to approach regionalism and regional identity in historical terms, we need to consider the interplay between the "imagined communities" of nationhood and local and regional differences issuing from actual social, cultural, and environmental conditions.[24]

By all of these measures, then, a "regional identity" constitutes a doubly slippery and inherently contradictory concept. On the one hand, the two primary components of the concept, "region" and "identity," are, in themselves, imprecise terms with indefinite boundaries. On the other, the notion of a "regional identity" implies an almost monolithic construct, one which has been effectively seized upon by many scholars who have studied American regions in general and the Mid-Atlantic region in particular.

If Zelinsky's "muted vagueness" about the collective personality is a hallmark of the Delaware Valley or lower Mid-Atlantic region, that vagueness also tends to be reflected in the work of historians and other scholars who have written about the area. Historical scholarship on the early Mid-Atlantic region is generally not as well-developed as that dealing with the Chesapeake and New England, in part because the heterogeneous Middle Colonies have been more difficult to characterize than their more easily defined neighbors to the north and south. Because social diversity and ethnic and religious pluralism *were* the area's defining characteristics, unifying themes and seminal works of scholarship are not easily isolated or identified. Still, despite this wide range of social, economic, and political factors, some historians have noted several regional commonalities, including increasing social and economic stratification, a growing dominance by merchant elites, and the burgeoning populations that characterized the more urbanized port towns of New York and Philadelphia.[25]

Historians have debated the case for such regional commonalities in the Mid-Atlantic, and several have argued both for and against the existence of a coherent "middle" region in the colonial and early national periods. To borrow from geographer Martyn Bowden, who has argued similarly against the broad regional coherence of colonial New England, some have conceptualized the Middle Colonies as a cohesive regional entity based only on "academic convention" or "invented tradition," simply because a number of colonies, and later states, happened to be contiguous.[26] Others have countered this assumption by maintaining that no real geographical or cultural basis existed for such a region. There were more important distinctions *within* the "region," they claim, but they have then attempted either to redraw Mid-Atlantic regional boundaries "more logically" or to justify the "real" existence of this "mythical" region based on interactive behavior.[27] These conflicting points of view summarize the essential contours of the debate. Does the middle region emerge as a definable entity, or is it, in fact, an applied construct that obfuscates rather than illuminates? And, if the middle region can be treated as a coherent unit, on what bases are its diverse parts similar?

Several other historians have either used the notion of regionalism as an overarching construct to argue for or against the existence of a coherent middle region or have implicitly acknowledged its actuality based on several criteria. Michael Zuckerman has suggested that the Middle Colonies, unlike the more easily delimited regions of New England or the Chesapeake, were defined by their very diversity, hence his coinage "the motley middle." Zuckerman challenged the exaggerated emphasis of the New England model of

American history and argued instead that the pluralistic Middle Colonies were the precursor to modern American diversity. Rather than following the models of Puritan New England or the Cavalier South, the newly emerging American society mirrored Pennsylvania pluralism, and was "its middle regions writ large."[28]

Jack P. Greene has viewed the Middle Colonies somewhat differently, defining them as a coherent entity based on several broad patterns. Although the ethnic diversity, variegated settlement patterns, and generally dual market orientation of this area created a pluralism that fostered much variety, Greene also saw broad similarities throughout the colonial period that seemed to argue strongly for considering it as one of six clearly delineated sociocultural regions in colonial Anglo-America. Although the Middle Colonies formed, more than any of these other regions, a pluralistic society exhibiting great variation, Greene also found a "common cultural core" resulting from a "steady process of social amalgamation" within a common physical and political environment.[29]

One of the most sweeping recent attempts to deal with the Mid-Atlantic in general and the Delaware Valley in particular as a coherent regional entity is David Hackett Fischer's *Albion's Seed: Four British Folkways in America*. Fischer's synthetic cultural history argues for the primacy of four different regional cultures in America based on voluntary migrations from four different English cultural centers, or hearths.[30] Fischer attributed some characteristics seen in much of the Delaware Valley region to the influence of English Quakers from the North Midlands and their sympathizers. Although churches of other denominations had outnumbered meetings of the Society of Friends by the mid-eighteenth century, the cultural imprint of the Quakers remained strong, and the tenets of Quakerism, including religious freedom, social pluralism, and a host of other attitudes, entered deeply into many aspects of Delaware Valley culture.[31] Fischer thus found that North Midland Quaker cultural imprints informed regional features as diverse as place names, speech ways, child-rearing ways, naming patterns, clothing, foodways, educational and family attitudes, and, of course, building practices.[32] *Albion's Seed*'s simplified portrayal of the area's built environment alone effectively demonstrates how basing an entire cultural analysis on such overarching and rigid regional constructs can sometimes confound rather than clarify, for the cultural landscapes of the early Delaware Valley remained far more diverse than Fischer's characterization suggests.[33]

While many historians such as David Hackett Fischer have utilized a regional construct to frame their interpretations of the Delaware Valley and the Mid-Atlantic within broader national studies, others have focused on specific portions of the area, often inferring or extrapolating generalizations about the whole region from their more specific studies. Barry Levy's work on early Quaker families focused specifically on the Radnor and Chester Meeting tracts just west of Philadelphia. While Levy argued that middling English Quakers who had emigrated to the Delaware Valley left a significant cultural imprint and established a distinctive American family ideology, he was not prepared to attribute the *whole* of Delaware Valley regional culture to the Quaker influence.[34] Stephanie Grauman Wolf's study of Germantown, Pennsylvania, in the eighteenth century provides another example of a locally based study that implicitly infers generalizations about the broader region based on a more geographically focused entity. Like Zuckerman, she argued for the suitability of Germantown and, by extension, the Mid-Atlantic, as a model for subsequent American development.[35] Sally Schwartz's examination of colonial Pennsylvania was somewhat more broadly based than Wolf's and Levy's works, but it still focused on a single colony. Schwartz, who maintained that Pennsylvania was far more pluralistic than the other middle colonies and had "no clearly dominant cultural group," brought localism into the foreground. Colonial Pennsylvania, according to Schwartz, constituted a culture that was "localistic in orientation, with allegiance not to a nation but to some subdivision of it," and this may have hindered the development of what she terms "ethnic consciousness" or group identity.[36]

One of the most influential regionally based studies for all historians of the Mid-Atlantic was written, not by a historian, but by a geographer. James T. Lemon's 1976 study of early southeastern Pennsylvania arguably stands as still the most important study of the entire southeastern Pennsylvania region. Lemon, who has also noted how difficult it is to define early American regional boundaries precisely, provided a thorough and synthetic look at a broad geographic area that encompassed or touched upon sixteen Pennsylvania counties and also dipped into adjacent parts of Pennsylvania, Delaware, Maryland, and New Jersey. Like Zuckerman and others, Lemon maintained that southeastern Pennsylvania constituted "the prototype of North American development" and that its lifestyle presaged the nineteenth-centry American mainstream. In his emphasis on the intertwined importance of environment and the decisions of its inhabitants in shaping life on the land in early Pennsylvania, Lemon argued for a kind of "liberal individualism" that was fostered by a sense of an "open" environment and which involved the maximization of benefits by individuals and their families without a great deal of attention to the common good. Like Greene, though, Lemon also deemphasized some aspects of the area's celebrated pluralism, maintaining, "Differences in customs and practices associated with national groups have . . . been misstated or exaggerated out of proportion to their significance."[37]

Several other geographically oriented studies of the Mid-Atlantic region have also laid important groundwork. The studies of Peter Wacker have focused on New Jersey, and his more recent contributions have also built upon the work of Lemon. Wacker found that New Jersey's settlement landscapes formed, not a hearth or a set of regions, but a complex cultural mosaic that did not necessarily always mirror distinct population regions. This mosaic was affected by factors as diverse as the initial plans of the proprietors, quality of soils, population density, boundary disputes, a possible folk-elite dichotomy, and the location of Native American trails.[38] Complicating the debate over regional commonalities even further, Wacker found that New Jersey stood apart from the other Middle Atlantic colonies in terms of its diverse geography and population, its methods of land allocation, and its emphasis on supplying agricultural and forest products for export to nearby ports and urban markets. In contrast with Lemon, Wacker suggested that New Jersey's several very distinct culture areas have also had a significant impact on land use and, by extension, cultural landscapes.[39]

Other geographical studies have discussed the entire Mid-Atlantic region and in somewhat broader terms. The geographer Wilbur Zelinsky, who termed the Middle Atlantic region "the Midland," has remarked that, of all of the American cultural regions, it is the "least conspicuous, either to outsiders or to its own inhabitants." Putting a different spin on the arguments of some historians, Zelinsky maintained that this relative inconspicuousness may actually argue for the region's centrality in the course of American development; in other words, it may be the region's defining characteristic.[40] Despite this acknowledged inconspicuousness of the region, Zelinsky attempted to delimit its boundaries. Within the broader Middle Atlantic region, he distinguished between an inland "inner zone" west and northwest of Philadelphia that he called the Pennsylvania subregion and a much more nondescript "outer zone." Still, Zelinsky conceded the mixed character of at least some parts of his Midland region and echoed Peter Wacker's findings for the mosaic-like quality of New Jersey. To Zelinsky, the areas to the west and north were "less clearly Midland in character, and much of central and southern New Jersey, except perhaps for a zone within a few miles of Philadelphia, is difficult to classify in any way, and is assigned to the Midland largely by default."[41]

While numerous historical and geographical studies have adopted a more sweeping approach to the lower Mid-Atlantic and Delaware Valley regions, most assessments of the region's built environment and related material culture have tended to focus more narrowly, either on relatively circumscribed locations or on the influence of particular ethnic or sectarian groups. Although most earlier studies concentrated on the documentation and description of specific buildings or groups of buildings within a particular geographic area, more recent studies typically seek fuller cultural interpretations.[42]

Several important architectural studies have centered specifically on Pennsylvania German regions and the building practices indigenous to those areas. G. Edwin Brumbaugh's 1930s work on Pennsylvania German architecture, while often downplayed by more recent architectural historians, nonetheless set an important precedent for much subsequent research. Brumbaugh, who advocated an archaeological approach to architectural analysis, concentrated on the domestic or domestically inspired "architectural monuments" of the Pennsylvania Germans. Equally significantly, he observed that the flow of architectural ideas was multidirectional (German to English and vice-versa) rather than simply unidirectional (English to German), and he argued, as did subsequent researchers, that Germany produced its own variation on the international Georgian style.[43]

Even more specifically focused studies of Pennsylvania German building practices were published several decades later. Many appeared in the pages of *Pennsylvania Folklife* in the 1960s. These include, for example, Robert C. Bucher's study of Pennsylvania German log dwellings, Art Lawton's examination of the relationship of Germanic measuring techniques to house form, and Henry Glassie's more theoretical analysis of the Continental plan house.[44]

More recently, Scott Swank used surviving buildings and the 1798 federal direct tax records to examine the process of acculturation in the architectural landscape across a broad swath of the Pennsylvania German heartland. Swank, who studied Pennsylvania German architecture in at least five Pennsylvania counties, determined that the architectural landscape in those regions experienced rapid change in the nineteenth century due to a widespread English Georgian architectural revolution.[45]

Several other studies of Pennsylvania German architecture have centered on more narrowly defined regions and the building practices indigenous to those particular areas. Philip Pendleton's work on the Oley Valley of Berks County, Pennsylvania, explores the architectural and cultural landscape of one of Pennsylvania's earliest settlements. Pendleton found that by the late eighteenth century, responses to American conditions coupled with contact between diverse settlement groups had produced a distinctive regional architecture in the Oley that not only combined old and new elements but also differed from European precedents. A distinct regional culture emerged that was shaped in part by the widespread adoption of the Pennsylvania agricultural system.[46] Similarly, Charles Bergengren's work on Schaefferstown, Pennsylvania, focuses on the architectural landscape of a single, predominantly Pennsylvania German area. Bergengren's research into changing house forms in this area suggests that changes in the "etiquette of entry," or the decision "to knock or not to knock" when entering the house of another, relate to changes in world view as well as house plans.[47]

While the Pennsylvania German imprint has garnered considerable attention, several location-specific studies have concentrated on other portions of the Delaware Valley's built environment. Some have dealt with southeastern Pennsylvania or portions of Delaware, while others have focused on all or part of New Jersey. Jack Michel's study of the built environment and material culture of early southeastern Pennsylvania tied socioeconomic status to material goods.[48] Bernard L. Herman's work on architecture and landscape in central and southern Delaware examined material and social landscapes as well as building and rebuilding cycles in those areas.[49] A collection of essays that accompanied a museum exhibit at the University of Delaware examined material life in Delaware in the early national period.[50] For New Jersey, Peter Wacker used documentary and architectural evidence to map house and barn types found throughout the state by the end of the eighteenth century. By comparing the distribution of building types to the known locations of ethnic and cultural groups in the state, Wacker postulated several distinct "culturogeographic regions and zones of mixture."[51] Even more regionally circumscribed are the writings of Paul Love, Alan Gowans, and Michael Chiarappa. Paul Love's 1955 study of the patterned brick houses in southern New Jersey focused primarily on documentation and laid much of the essential groundwork for Alan Gowans's and Michael Chiarappa's subsequent work on this important group of buildings. Alan Gowans maintained that southwestern New Jersey's patterned brick houses were the product of a Homeric aristocracy, while Michael Chiarappa's study of the region's Quaker brick artisans suggested that Quaker pattern brickwork constituted a territorial response to Quakerism's dwindling presence in the region at the end of the eighteenth century.[52]

Finally, of all the studies dealing in any way with the Delaware Valley built environment, Henry Glassie's contributions stand out, both for the depth of their cultural analysis and for the breadth of their regional approach. Glassie examined the process of cultural change in the Delaware Valley through buildings and farm plans. He also delineated a general Mid-Atlantic folk culture region—arguably the most important of all U.S. regions because it influenced both the north and the south—that included southeastern and south central Pennsylvania, southern New Jersey, most of Delaware, and all of northern Maryland. It was this area's mostly dual market orientation as well as its inherent cultural pluralism that set it apart. That tradition of pluralism carried over into the material world, for, as Glassie was quick to note, in the Mid-Atlantic, the common pattern of expression in material objects involved one of several choices: the blending of related European traditions, an emphasis on a single tradition, or the emergence of an entirely new culture that was Pennsylvanian rather than simply English or German.[53]

In short, while many historical studies of the Delaware Valley and the lower Mid-Atlantic have tended to use regionalism as an overarching interpretive construct, most of the material culture literature has tended to focus on more circumscribed geographic locations or on particular ethnic, sectarian, or social groups. The more geographically oriented studies seem to fall somewhere in between these two approaches. This essay aims to close some of the gaps within the literature on the Delaware Valley in particular and within the literature on regional and national identity in general. By examining and contextualizing several specific Delaware Valley landscapes, this study explores the significance of localism in the formation of a broader regional self-consciousness. Regions, as some scholars have suggested, are both actual and conceptual places; they are based on real as well as imagined distinctions; and they are constantly being re-formed, not only in relation to other places, but also in response to historical and cultural change.

→‖← →‖← →‖←

When geographers, journalists, linguists, and folklorists have attempted to map the boundaries of American culture and vernacular regions, their characterizations of the Delaware Valley usually more or less overlap in several key areas. Most typically agree on the location of an important cultural core or source area centered in southeastern Pennsylvania. Most also tend to concur that some sort of east-west boundary running through the midpoint of the Delmarva Peninsula divides the Delaware Valley from points south.[54] Still, despite these broad general points of agreement and despite the acknowledged importance of the lower Delaware Valley as a major cultural hearth, this area even today lacks the degree of definition that characterizes neighboring regions.

If "a muted vagueness about their collective personality" pervades the consciousness of inhabitants in the Pennsylvania culture area today, to what extent did such regional self-awareness exist in the past?[55] In the mental maps of historical actors and observers, how was this region perceived and experienced? Period accounts leave little doubt as to the highly localized nature of this region's landscape. Travelers through the area, encountering their surroundings at carriage and stagecoach pace, were intensely aware of how the landscape shifted as they moved through it. When the duc de la Rochefoucauld-Liancourt traveled north and west of Philadelphia in the 1790s, he observed how the construction of the buildings he saw began to change: "log-houses, constructed of trunks of trees, laid one upon another, the interstices of which are filled with clay" had begun to disappear, replaced by "framed houses . . . properly hewn and shaped, and covered with boards" as well as buildings of stone and brick.[56] Théophile Cazenove observed how the soil quality shifted from "inferior red gravel" to well-watered limestone land as he passed through York County, Pennsylvania.[57] Isaac Weld remarked that, on the approach to Philadelphia by water, the left bank of the Delaware was "much cleared" and "interspersed with numberless neat farm-houses, with villages and towns . . . some parts cultivated down to the very edge of the water," but the right bank, on the New Jersey side, remained "thickly wooded, even as far as the city."[58] And as he neared Conestoga, Pennsylvania from the east, Thomas Anburey wrote, "A new country presents itself, extremely well-cultivated and inhabited," with roads lined with log and occasionally stone farmhouses.[59]

As contemporary observers traveled through the Delaware Valley, they encountered, not the broad regional commonalities posited by modern-day scholars, but a congeries of localized landscapes. The differences they mapped, while far more muted today, were nevertheless real ones, at least to them. Shaped by factors ranging from soil types, settlement patterns, and sectarian imprints to economic, sociocultural, and ethnic differences, these very land-scape distinctions defined the Delaware Valley in the early national period. Yet, as America entered the first few decades of the nineteenth century, many of these distinctions were being redefined. The landscape was beginning to be reworked, and even as some of these observers were writing, the cultural landscapes they encountered were beginning to change.

# CHAPTER 2

# Ethnic Perceptions, Ethnic Landscapes

As Johann Schoepf traversed the Pennsylvania countryside north of Philadelphia at the end of the eighteenth century, he reflected on the visible marks that ethnicity had already made upon the early national landscape: "From the exterior appearance, especially the plan of the chimneys, it could be pretty certainly guessed whether the house was that of a German or of an English family . . . ," he observed; "if of one chimney only, placed in the middle, the house should be a German's and furnished with stoves, the smoke from each led into one flue and so taken off; if of two chimneys, one at each gable end there should be fire places, after the English plan."[1]

Houses, according to Schoepf's bald statement, spoke volumes about difference—cultural behavior was inextricably intertwined with, and expressed by, the material world. "Where a German settles," he wrote, "there commonly are seen industry and economy . . . his house is better-built and warmer, his land is better fenced, he has a better garden, and his stabling is especially superior; everything about his farm shows order and good management in all that concerns the care of the land."[2] Blunt though Schoepf's comments appear, his observations were not unusual. Contemporary correspondents echoed his observations, describing a world in which differences were clearly mapped by objects as well as actions, a landscape in which memory, tradition, and shared experience had already fused buildings, fences, and fields into un-mistakable cultural boundaries.

At the close of the eighteenth century, in an American countryside that was fast becoming resonant with symbols evoking national ideals of property, pros-perity, and agrarian conquest, the notion of distinct and local cultural bound-aries seems oddly contradictory.[3] Yet comments about perceptible cultural dif-ferences appeared frequently in contemporary accounts, especially those from the Pennsylvania backcountry, a place where settlement by a mosaic of different ethnic and sectarian groups that included English Quakers, Ger-mans, Welsh, and Scots-Irish had already created a pluralism that observers found remarkable. "In traveling through Pennsylvania," wrote the British of-ficer Thomas Anburey in 1789, "you meet with people of almost every differ-ent persuasion of religion that exists; in short, the diversity of religions, na-tions, and languages is astonishing."[4] Comments about the exceptionalism of the German population in this area—especially in terms of their industry, their frugality, their self-imposed isolation, and their farming skills—were firmly entrenched stereotypes in the popular literature by the end of the eigh-teenth century. While such remarks often centered on language differences, manners, and cultural practices, many also focused on the material world and the ways in which the land itself had been shaped.

References to such distinctions, common by the early national period and prevailing through the early nineteenth century, implied clear and dis-cernible cultural boundaries. But were these boundaries really as visible as contemporary writers claimed? To what extent did these correspondents pro-ject their own attitudes about cultural otherness—attitudes that may have been shaped just as much by the larger cultural process of forming a national identity—onto the early national landscape? As the new nation matured, were these observers instead engaged in a much broader, self-conscious process of reinventing local, regional, and national identities? Were the cultural borders they encountered, in fact, shifting social constructions that could be read dif-ferently by observers from different vantage points?[5] And, how did ethnically distinct groups such as the Pennsylvania Germans respond to the inevitable pressure to assimilate at the turn of the century, and to what extent was this drift toward assimilation a reciprocal process, a give and take across the in-creasingly permeable boundaries of several cultures? More to the point, can we actually discern the distinctions that these eighteenth-century correspondents described in the documentary and material records, or are such distinctions in the landscape more shifting and ephemeral, more deeply intertwined with the textures of everyday experience, what Dell Upton has called "invisible" material culture—the smells, sounds, fleeting sights, and emotional responses that are the incidental byproducts of cultural action but that nevertheless affect our responses to the physical world?[6]

James Lemon, using contemporary travel accounts in conjunction with census records, soil surveys, local tax lists, and agricultural journals, examined some of southeastern Pennsylvania's more persistent ethnic stereotypes in light of quantifiable agricultural practices to determine whether Pennsylvania German farming practices warranted the high praise they had historically garnered. By mapping soil types, land use, and animal husbandry practices, Lemon discovered that the most deeply embedded ethnic stereotypes were *not* founded on measurable agricultural practices but arose from a combination of inherited English attitudes and prevailing ideals. Lemon argued that success in agriculture was determined by overall economic prosperity rather than by ethnic origin, and he speculated on reasons for the relative lack of egalitarianism toward all national groups in the historical literature. Noting that references to "English" Pennsylvanians or "English" farming practices in America are comparatively rare, he suggested that "we have tended to look through 'English' eyes at our society," distinguishing minority groups more sharply than the practices of the dominant group.[7] Lemon's discoveries are significant and suggestive, but his work focuses primarily on the agricultural rather than the architectural environment. And although most period commentators remarked on differences in land stewardship practices as well as manners, many also described perceptible differences in houses, fences, and barns—the built stuff of the material world. This discussion returns to the material world by examining the landscape through the lenses of several different forms of evidence, including contemporary accounts such as Schoepf's, the built landscape itself, and period tax records. One tax in particular, the 1798 federal direct tax, offers us the opportunity to look more closely at patterns in land use and building practices in the Pennsylvania backcountry.

The 1798 federal direct tax, which is often called the "glass tax," was the nation's first steeply progressive real estate tax. Devised by Congress to raise income to support a standing military, it focused on land, buildings, and slaves. Congress authorized state-appointed assessors and their assistants to

record specific details about the built landscape, and they used a complex re-
porting system that included several different lists, called "particular lists." Assessors recorded not only the names of property owners and the value of their
holdings but also listed a wealth of additional information, including building dimensions, construction materials, total acreage, whether the property
was tenanted or not, the names of the tenants, information on outbuildings,
the number and size of windows present (hence the label "glass tax"), and total assessed value. The information in this tax record is especially useful for
studying early American landscape organization, because it suggests a great
deal about settlement patterns, economic activity, tenancy rates, and early industrialization—and ethnicity. Moreover, because the tax required a uniform
level of information on all real property, it permits us to contextualize individual surviving buildings in both historical and physical terms. Surviving
physical context, as any student of the built environment soon discovers, is a
rare commodity. Buildings of all types often survive today in isolation, without
their original surroundings: while the dwellings that once formed the focal
points of many early farmsteads often survive, the outbuildings that typically
clustered nearby usually do not. Similarly, historically "significant" buildings
in urban areas often stand in sharp contrast to their dramatically altered
 present-day environments. The 1798 tax thus allows us to determine, through
comparison with the aggregate, what was typical as well as what was rare.

This comprehensiveness makes the direct tax a reasonably egalitarian tool
for examining the built environment. And although the descriptions themselves lack uniformity from one geographic area to another, the tax was designed to be evenly assessed everywhere, from the most ethnically homogeneous German- or English-settled townships to subregions that had already
become remarkably heterogeneous.[8] This uniformity allows us to sidestep the
historical tendency to view the landscape through the "English eyes" of the
historical majority and look more critically at landscapes that appeared rarely
or were described in biased fashion in contemporary accounts. Instead, because tax assessors were locally appointed, we are permitted a view of the built
world through an insider's eyes.

Despite this superficial uniformity, though, the aggregate picture that
emerges from the 1798 tax in the Delaware Valley is one of a patchwork of localized landscapes, each of which manifested its own distinctive idiosyncrasies.
Settlement patterns, building types, economic activity, and overall landscape

organization varied widely within the broader Delaware Valley region and even from one southeastern Pennsylvania township to another. Thus, the systematized uniformity of the tax paradoxically both underscores and belies the widespread localism of the early national countryside.

With its comprehensive assessment of the architectural landscape, then, the 1798 tax provides a useful source for examining such questions of ethnicity. By using the tax to investigate several Pennsylvania townships defined by a strong and recognizable ethnic character and comparing them to less ethnically distinctive areas, we can begin to gauge the distinctiveness of and interconnectedness between such landscapes. Although tax assessors often failed to record the most obvious ethnic differences that travelers so frequently noted, they did provide a relatively unbiased record of *all* of the buildings they saw, from the most magnificent mansion to the lowliest cornhouse. By utilizing the descriptive information contained in these tax lists in conjunction with contemporary descriptive accounts, other documentary sources like later tax lists and probate inventories, and surviving buildings, we can explore expressions of ethnicity in the built environment and begin to reconstruct the material context of Schoepf and his contemporaries.

While both tax assessments and contemporary travel accounts are essentially descriptive documents that record earlier landscapes, and both are subjective texts that were ultimately controlled by their authors' perceptions and values, they are nonetheless two different types of documents generated for very different purposes. Tax assessors viewed houses, barns, and fields as property—these objects constituted tangible economic assets with dimensions to be inventoried, windows to be counted, construction materials to be noted, and cash values to be tabulated. Unlike travel writers and contemporary observers, tax assessors also itemized and described most architecture equally; even less "public" buildings such as springhouses and cornhouses failed to escape mention, because they had economic value. But tax lists, despite their detail, seldom capture texture and evocative power. Contemporary observers, on the other hand, seized upon manners, customs, appearances, and differences, and sometimes expounded at great length on what they saw and how they felt about it, but their accounts often lack the unvarnished precision and thoroughness of the 1798 direct tax records. Still, both types of document provide useful starting points. By observing the ways in which these two varieties of description intersect or diverge, we can explore the boundaries of the ethnic landscapes that Johann Schoepf and his contemporaries described and can begin to unravel the difference between the textual and the physical, between the landscape of perception and that of experience.

Returning to the contemporary accounts, we find that observations such as Schoepf's were common, especially in reference to the Pennsylvania Germans. To Schoepf, they lived "thriftily, often badly," maintaining the traditions of industry and extreme frugality that they had learned in the old country, even when their English neighbors provided a "better" example. "There is wanting among them the simple unaffected neatness of the English settlers, who make it a point, as far as they are able, to live seemly, in a well-furnished house, in every way as comports with the *gentleman*," he wrote. Schoepf scorned the Pennsylvania Germans' tightfistedness and bemoaned their cultural insularity. "To use their gains for allowable pleasures, augmenting the agreeableness of life, this very few of them have learned to do, and others with a bad grace," he observed. "The lucre is stuck away in old stockings or puncheon chests until opportunity offers to buy more land which is the chief object of their desires." The typical Pennsylvania German interior, as described by Schoepf, made cultural differences manifest in material form: "A great four-cornered stove, a table in the corner with benches fastened to the wall, everything daubed with red, and above, a shelf with the universal German farmer's library: the Almanack, and Song-book, a small 'Garden of Paradise,' Habermann, and the Bible. It is in vain to look for other books."[9]

Theophile Cazenove, traveling through the Pennsylvania backcountry in 1794, concurred, praising the well-tended landscape but criticizing the frugality of the Pennsylvania Germans: "they are thrifty to the point of avarice; to keep seems to be their great passion; they live on potatoes, and buckwheat cakes instead of bread." While Cazenove found the exteriors of some German houses appealing, he thought the interiors to be "almost unfurnished." Sparse table furnishings bespoke lives lacking comfort and civility, sleeping quarters rarely provided anything more than beds and a margin of privacy, and most dwellings contained little more than a forlorn jumble of food and domestic flotsam: "For downstairs rooms, a kitchen and a large room with the farmer's bed and the cradle, and where the family stays all the time; apples and pears drying on the stove, a bad little mirror, a walnut bureau—a table—sometimes a clock." In the largest, most formal room of the house he saw "an immense stove on which the dishes are still standing; potatoes and turnips on the floor; beds are generally without curtains, no mirrors, nor good chairs, nor good tables and wardrobes." Cazenove, who likened wealthy Pennsylvania Germans to farmers elsewhere of more modest means, chided them for their cultural parochialism. "You always feel like settling in the country when you see the excellent ground and the charm of the country, and also the advantage of farming," he wrote, "but you lose courage when you realize the total lack of education of the farmers, and that it is absolutely necessary to live to yourself, if you have any education, knowledge, and feeling."[10] The surveyor James Whitelaw agreed, praising the appearance of the German-settled countryside but characterizing the inhabitants as tight-fisted and land-hungry. "The province of Pennsylvania seems the most desireable to live in of any place we have yet seen." But the best land—fertile soil that was close to transportation and markets—was either too expensive to purchase or already taken by Germans. "Here the people are kind and discreet except the dutch or Germans who inhabit the best lands in this province," Whitelaw complained. The Germans were "a set of people that mind nothing of gayety but live niggardly and gather together money as fast as they can without having any intercourse with any body but among themselves."[11]

Similar characterizations recurred throughout contemporary literature. As early as 1753, a traveler in the Pennsylvania backcountry, remarking on the different priorities the Germans accorded to their dwellings and their agricultural spaces, wrote, "It is pretty to behold our back Settlements, where the barns are as large as pallaces, while the Owners live in log hutts; a sign tho' of thriving farmers."[12] Writing in the 1790s, the duc de la Rochefoucauld-Liancourt summed up the same Pennsylvania German landscape characteristics succinctly: "The houses are small, and kept in very bad order; the barns are large, and in very good repair." West of Lancaster he observed a similar trend. "We met with scarcely any but log-houses; every where we observe German farms, small houses, and large barns."[13] Benjamin Rush, in a lengthy article printed in the *Columbian Magazine* of January 1789 offered one of the most famous commentaries of all, listing a series of landscape characteristics that were directly attributable to the Germans and neatly summarizing the essence of the stereotype. To Rush, the differences between the German cultural landscape and the surrounding region were visible and striking: "A German farm may be distinguished from the farms of the other citizens of the state, by the superior size of their barns; the plain, but compact form of their houses; the height of their enclosures; the extent of their orchards; the fertility of their fields; the luxuriance of their meadows, and a general appearance

of plenty and neatness in every thing that belongs to them."[14] The image of
Pennsylvania German farmers as frugal, uncultured, and unbudging remained in 1836, when Tyrone Power, an Irish actor, traveled by stage from
Philadelphia to Pittsburgh and wrote: "The country through which we rode
was under excellent cultivation; the barns attached to the road-side houses
were all large, brick-built, and in the neatest condition." But, he continued,
"In his piggery of a residence and his palace of a barn, in his wagon, his oxen,
his pipe, his person and physiognomy," the third-generation Pennsylvania
German farmer remained unchanged.[15]

Commentators thus consistently noted several characteristics of the German-settled backcountry. Fine, carefully tended land, large and well-built
barns that formed the centerpiece of every farm, and houses that were modest by comparison—these were the physical manifestations of the frugality, industry, and parochialism that observers celebrated or bemoaned; these were
the material boundaries of Pennsylvania's German culture. But what were the
catalysts for such reactions? What was it about the Pennsylvania Germans that
provoked responses that were strong and consistent enough to generate such
lasting stereotypes? How did period commentators view ethnic distinctions in
relation to the boundaries of their own culture, and how did they compre-
hend the notion of ethnicity itself?

Ethnicity, in particular, is a slippery notion characterized by shifting meanings and changing levels of significance over time. Several definitions of ethnicity, according to Kathleen Conzen and other historians, have dominated
most discussions of immigrant adaptation. One has tended to emphasize its
primordial aspects. Through this lens, ethnicity is genealogy. It constitutes a
unity that emanates from the shared ancestry, culture, and "group identity" of
human beings and is "ancient, unchanging, inherent in a group's blood, soul,
or misty past." An alternative conception of ethnicity, sometimes characterized as "emergent ethnicity," downplays this notion of primordial solidarity
and instead views it as an organizing construct for interest groups. Ethnicity
by this definition serves as a way to mobilize and empower certain subpopu-
lations, enabling them to compete for socioeconomic position.[16] More recently, Conzen, Dell Upton, and other scholars have argued instead for a more
complex view of ethnicity, as both an invention and a shifting construct, a fluid
cultural edifice that metamorphoses, disappears, and reappears depending
upon changing relationships within the smaller ethnic group itself and with
the surrounding society. According to this notion of "invented" or "dynamic"

ethnicity, group boundaries must constantly be redrawn and renegotiated, and the most expressive symbols of ethnic group identity must be redefined. Ethnicity, then, through this lens, is constantly mutating; it is "a role played for the benefit of others."[17]

This more fluid definition of ethnicity proves especially useful when examining the period of change immediately following the Revolution—that period when symbols of nationhood were developing and attaining intense cultural importance. For, as Wilbur Zelinsky has argued, America was "symbolically impoverished" before 1776, "lacking its own music, literature, art, native costume, or cuisine, and homegrown heroes. . . . There was no capital city as yet, no well-entrenched ruling class, nor any conventional military caste or even a name for the country that carried much emotional force." Like national symbols, ethnic identities during the post-Revolutionary period were undergoing a process of definition and redefinition, and any consciousness of a general American ethnicity was barely formed by 1775.[18]

Still, if ethnicity is, in fact, an invention and a shifting cultural construct, most commentators during this period were probably too close to it to see it that way. Contemporary observers most often tended to view ethnicity as a primordial thing—something that was innate and based on ancestry, something that, like their painted wooden chests, traditional songs, and religious customs, the immigrant groups brought with them and often sought to keep. Few observers, if any, seem to have stepped back far enough to consider ethnicity any differently. When the duc de la Rochefoucauld-Liancourt passed through the vicinity of Germantown, Pennsylvania, in the 1790s, he betrayed his own personal opinions when he reported that "the [mostly Germanic] inhabitants are by no means intelligent, and they are particularly averse to leave their old customs for a new method which might be better; but they are industrious, and their assiduity to labour counteracts, in some measure, their repugnance to all improvement."[19] Henry Fearon described the bulk of Pennsylvania farmers similarly: they were often rich and mostly "Germans, or of German descent. They are excellent practical farmers, very industrious, very mercenary, and very ignorant."[20] Other assessments of the Pennsylvania Germans also characterized them as "avaricious, egoistic, and . . . willing to oblige you only insomuch as their interest urges them to do it." Their sole purpose was to make money: "When they came to settle among us, it was neither to acquire knowledge nor to disseminate it, but to make a fortune." The Pennsylvania German love of wealth was particularly notable to many observers, and

Fearon, like some others, blamed the Pennsylvania Germans' failure to assimilate on their miserly behavior. "Their favorite passion for money, and their unfortunate condition," he wrote, "have kept them in the original state of ignorance of their fathers."[21]

This focus on money was important to many contemporary observers. For, if most saw ethnicity as innate, as something that was derived from ancestral ties, they also tended to view ethnic distinctions through a wealth-biased or class-biased lens. In particular, perceptions of the rich but parsimonious "Dutch" (as the Germans were often called) were often colored by the observers' notions of what "success" should mean and how wealth should be used. Successful Pennsylvania German farmers who valued money yet seldom cared to display their wealth conspicuously challenged prevailing expectations with their values. More to the point, many observers viewed this apparent inversion of ideals as somewhat problematic and, perhaps, a bit threatening. Theophile Cazenove, who constantly marveled at the disjuncture between the wealth of most Pennsylvania German farmers and their actual physical circumstances, complained "if the [Pennsylvania German] farmers liked money less, they would surround themselves with more conveniences and live in plenty."[22] "The German farmers," Cazenove wrote, "manufacture coarse woolen material for coats, shirts, etc., and all their shirt-linens. They buy only their best clothes for Sunday and not many of these" due to their excessive thrift. Still, Cazenove acknowledged that the Germans had their priorities. "They deny themselves anything costly; but when there is snow, they haunt the taverns. They are remarkably obstinate and ignorant."[23] To Cazenove and other observers, it seemed, wealth should clearly be translated into creature comforts and material goods. Pennsylvania Germans who lived parsimoniously not only flaunted prevailing ideals; they effectively redefined success on their own terms. Thus, while the Germans resembled the cultural mainstream in their pursuit of wealth and property, it was their failure to value material comforts in the same way others did that made them appear so distinctive.

Some commentators seem to have conflated ethnic and status distinctions. In such cases, while the reaction to distinctive behaviors is clear, the real basis for these observers' reactions to the "foreignness" they describe is not clear, and this presents a tangle of conflicting possibilities. Francis Baily, the son of a British banker, traveled through North America in 1796 and 1797. Baily's telling description of an encounter with a Pennsylvania German family near

Chambersburg betrays a number of cultural expectations that seem to derive primarily from his own status-based perceptions:

> About thirteen miles from Chambersburgh, which we left in the afternoon, is a place called the *Mill*, which is kept by some Dutchmen. We understood it was a tavern, but were disappointed; however, as it was now dark, and no tavern on the road for some distance, we were under the necessity of begging a lodging here, which was granted us at last with the greatest reluctance. Here we had rather an unfavourable specimen of Dutch manners. We were *kindly* directed to take our horses to the stables, and take care of them ourselves, which we accordingly did; and, returning to the house, I was witness to a kind of meal I had never before experienced. First of all, some sour milk was warmed up and placed on the table. This at any other time would probably have made us sick; but having fasted nearly the whole day, and seeing no appearance of anything else likely to succeed it, we devoured it very soon; particularly as the whole family (of which there were seven or eight) partook of it likewise; all of us sitting around *one* large bowl, and dipping our spoons in one after another. When this was finished a dish of stewed pork was served up, accompanied with some hot pickled cabbage, called in this part of the country "warm slaw." This was devoured in the same hoggish manner, every one trying to help himself first, and two or three eating off the same plate, and all in the midst of filth and dirt. After this was removed, a large bowl of cold milk and bread was put on the table, which we partook of in the same manner as the first dish, and in the same disorder. The spoons were immediately taken out of the greasy pork dish, and (having been just cleaned by passing through the *mouth*) were put into the milk; and that, with all the *sang froid* necessarily attending such habitual nastiness. Our *table*, which was none of the cleanest (for as to *cloth*, they had none in the house), was placed in the middle of the room, which appeared to me to be the receptacle of all the filth and rubbish of the house; and a fine large fire, which blazed at one end, served us instead of a candle.
>
> Wishing to go to bed as soon as possible (though, by the by, we did not expect that our accommodations would be any of the most agreeable) we requested to be shown to our room, when lo! we were ushered up a *ladder*, into a dirty place, where a little hole in the wall served for a window, and where there were four or five beds as dirty as need be. These beds did not consist (as most beds do) of blankets, sheets, &c., but were truly in the Dutch style, being literally nothing more than one feather bed placed on another, between which we

were to creep and lie down. The man, after showing us this our place of destination, took the candle away, and left us to get in how we could, which we found some difficulty in doing at first; however, after having accomplished it, we slept very soundly till morning, when we found we had passed the night amongst the whole family, men, women, and children, who had occupied the other beds, and who had come up after we had been asleep. We got up early in the morning from this inhospitable and filthy place, and, saddling our horses, pursued our journey.[24]

Baily's obvious repugnance at the lack of manners, cleanliness, and personal privacy in this Pennsylvania German household is equally matched by his shock at having to force down ethnically unique foods such as stewed pork and warm slaw and having to sleep between "Dutch-style" featherbeds that did not resemble "most beds." More revealingly, while his reactions to the ethnic distinctiveness of his hosts seem to be derived as much from his own societal expectations of what "proper" behavior and accommodations for one of his station should consist of, he tends to fuse the two. In this way the modest surroundings and unschooled manners of this Pennsylvania German family are also seen as a product of their ethnic origins as well as an aspect of their ethnic distinctiveness.

If contemporary accounts sometimes freely mixed status and ethnic distinctions when referring to the Pennsylvania Germans, they also referred to the distinguishing characteristics of English culture only rarely, as James Lemon has noted, and period observers routinely viewed other ethnic groups through "English eyes." Such myopia evidently has a long history. Conzen et al. maintain that perceptions of ethnic differences between colonial settlement groups usually happened selectively, because only some kinds of foreignness were initially seen as problematic. Thus, the distinctive behaviors of English immigrants were never really singled out the same way those of other immigrant groups were, because "the English had no ethnicity in American eyes." Still, although most Americans may have been aware of ethnic differences, the majority tended not to worry about them, because they had faith that the Americanizing process would eventually erase most distinctions. Conzen et al. argue that the idea of ethnicity gained resonance only toward the mid-nineteenth century, when Americans began to view themselves and their society differently and when they began to worry more about distinctions within that society. Nonetheless, the trajectory of this shift was gradual and uneven. Perceptions of ethnic differences began to solidify "first in religio-ethnic, then increasingly in purely ethnic, terms."[25]

Period comments support this point. Although contemporary observers tended to see all ethnic groups through English eyes, they also often assumed that ethnic distinctions would blur once exposed to the broader Anglo-American culture. Some period accounts even reveal observers' suppositions about the predicted trajectory and timing of acculturation. For example, Isaac Weld, an Irishman traveling in America in the 1790s, described the Germans as skilled and religious farmers, "most valuable citizens" and "a quiet, sober, and industrious set of people." According to Weld, Germans also exhibited a distinctive attitude toward the land. While a German preferred to purchase and work land near his relatives and remain in one place, Weld remarked, most Americans presented "a roving disposition" and continually moved from place to place in search of cheap land and greater wealth. Nonetheless, he saw one similarity between Germans and Americans in their love of money. "The American, however, does not change about from place to place in this manner merely to gratify a wandering disposition; in every change he hopes to make money," he remarked. "By the desire of making money, both the Germans and Americans of every class and description, are actuated in all their movements."[26] Despite this mutual desire for wealth, though, because the Germans constituted "a plodding race of men, wholly intent upon their own business, and indifferent about that of others," and because they typically settled near one another in large groups, they tended to retain their native customs and language. Weld, who expected that acculturation would only be a matter of time and exposure, noted that "the Germans and their descendants differ widely from the Americans, that is, from the descendants of the English, Scotch, Irish, and other nations, who from having lived in the country for many generations, and from having mingled together, now form one people, whose manners and habits are very much the same."[27]

Similarly, Theophile Cazenove expected that the Pennsylvania Germans would eventually adapt to English ways. Yet his remark about the discrepancy between the attractive "English" exteriors of some German farmers' houses and their humble and sparsely furnished "German" interiors also betrays his assumption that this transformation would happen slowly, and might even take a generation or two. Like Weld, Cazenove maintained that cultural change would occur only when the Pennsylvania Germans, who "do not know any better," were exposed for a prolonged period to non-German ways. And, according to Cazenove, such cultural change would follow a forseeable pattern in the material world. "You notice especially," he wrote, "the clothing of the German farmers and their wives who have an opportunity to see other examples than their father's and mother's; they have English or American clothing, and from clothes it will pass to house-furnishings, etc." To Cazenove, then, the path of acculturation was clear: with their clothing as with their houses, Pennsylvania German immigrants would shed their Germanness gradually. More significantly, this transformation would occur in a predictable sequence, starting from the outside in.[28]

⁂ ⁂ ⁂

Although commentators tended to lump them together into a single ethnic group, by 1798, the Pennsylvania German "community" was, in fact, an extraordinarily diverse group, an uneven assortment of more recent German-speaking immigrants and the offspring of earlier settlers who had emigrated from many different regions. Most Pennsylvania Germans felt local rather than national allegiances. Although united by some commonalities, such as language and shared local traditions regarding family, work, and inheritance, German settlements in Pennsylvania were also fractured by diverse religious affiliations and backgrounds. The composition of the incoming immigrant population also changed noticeably over time. While German-speaking settlements in the early eighteenth century were a somewhat more homogeneous lot because they consisted largely of families with money or access to transatlantic support networks, diversity among new immigrants increased significantly around midcentury. In fact, at the peak of the German in-migration in the 1740s and 1750s, the incoming families grew poorer overall, and the new immigrants included a higher proportion of single people. After the Seven Years' War, German immigration declined, and the composition of this population had shifted again, to include mostly young men traveling by themselves. Still, despite this obvious diversity, most observers persisted in viewing this mixed community through English eyes. In the words of one historian, "only their British neighbors could describe the Germans indiscriminately and monolithically as one group."[29]

Also diverse were the ways in which the Pennsylvania German "community" interacted with, changed, and was changed by the broader and equally diverse Anglo-American culture. While Cazenove, Weld, and other contemporaries predicted that the Pennsylvania German element *would* eventually adapt to the cultural mainstream, the ways in which they actually adapted or resisted varied. In his study of Pennsylvania German material culture, Scott Swank argued that Pennsylvania Germans attempting to forge an identity in the American environment essentially followed three basic trajectories: total assimilation, which usually meant the abandonment of old affiliations; outright rejection, which required a significant effort to maintain traditions and resist the pressure to assimilate; and controlled acculturation, which usually meant a gradual but steady movement toward assimilation.[30] The kind of adaptation he refers to would progress from ethnic solidarity to acculturation, proceeding along a somewhat linear path, and assimilation would be more or less unidirectional. But while cultural continuities and discontinuities between the Old World and the New World were certainly present, and varying degrees of assimilation, gradual acculturation, and resistance undoubtedly occurred, culture change is far more dynamic and complex than such a model might allow. Old solidarities can give way to new and different identities, and linear explanations for cultural transformation can fail to capture the complexity of the change as well as the uneven and reciprocal nature of the interaction. For, despite Cazenove's contention that Pennsylvania Germans exposed to mainstream ways would gradually adopt a more "English" manner of dressing and furnishing their houses, their material culture was also known to change in the opposite direction, and at least a few Englishmen were known to have built "German-style" dwellings to please their German wives.[31] In Pennsylvania as elsewhere, cultural change seldom proceeded in a linear fashion, and "minority" cultures such as the Pennsylvania Germans also exerted profound and lasting effects on the cultural mainstream.

If we accept the notion of dynamic or invented ethnicity, if ethnicity constitutes a boundary-making process and "a role played for the benefit of others," then material objects and the identities so often coupled with them may also constitute a commodification of those roles, a choice rather than a fact. As Dell Upton has noted, ethnicity is not only formed and reformed depending upon circumstance, but it is also "a creolized identity and a highly volatile one."[32] By this definition, then, ethnicity is under constant scrutiny and subject to constant change. Thus, as we consider how ethnic identities are expressed in the material world, it may be more useful to approach cultural transformation in terms of creolization rather than movement from ethnic solidarity to assimilation or acculturation. Adapted from a linguistic model, the notion of creolization—defined here as a syncretic process of cultural interaction, negotiation, and convergence that ultimately gives rise to a new and different culture—will prove especially useful as we confront the complexity of the Pennsylvania built environment in the late eighteenth and early nineteenth centuries.[33]

It was the generations immediately following the Revolution that probably confronted the process of identity formation at its most complex. "The Revolutionary generation," argue Kathleen Conzen et al., "faced two fundamental problems of self identity: the need to differentiate themselves from Britain and the need to draw together states whose populations had very different cultural traditions and national origins. Nationality defined as culture and descent would have served neither purpose well."[34] It was also during this period that many Pennsylvania Germans would, as Cazenove had predicted, begin to shed their German ways, but many would also begin to forge their cultural identities anew. Still, no matter what patterns the cultural transformations in the Pennsylvania backcountry followed, period accounts suggest that those changes were hardly seamless. The cultural transition left visible signs in its wake, signs that contemporary correspondents and assessors collecting information for the 1798 direct tax also often noted.

When the direct tax was assessed at the close of the eighteenth century, Lancaster County, Pennsylvania, remained a largely agricultural landscape settled by a mixture of Reformed and Lutheran Germans, British Quakers, Scots-Irish, Welsh, and French Huguenots. The soil was naturally fertile, and farming was central to the lives of most of the county's residents. The subsistence farming of the earliest settlers had gradually given way to a mixture of subsistence and commercial farming, and this shift in emphasis in turn fostered commercial and industrial activity and the development of transportation networks. Milling was the earliest industry to develop in the county. Grist and flour mills were soon followed by saw, rope, and paper mills, all of which necessitated a growing transportation infrastructure. Because of the early development of the town of Lancaster as a major inland market town, the roads that connected such mills and farms to their broader markets were also fairly well developed by the end of the eighteenth century. North-south trade routes linked the town to Maryland and north-central Pennsylvania. In addition, the King's Highway, the major artery connecting Philadelphia and Lancaster, had already been built by 1730 and by 1792 was such a significant route that it

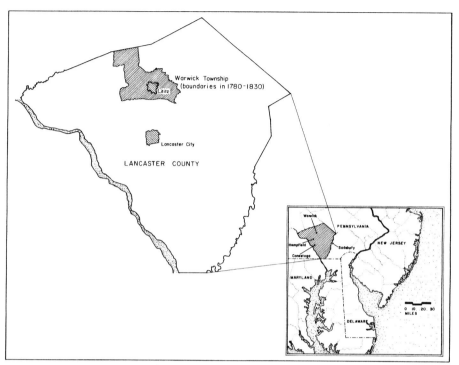

FIG. 2.1. Warwick Township, Lancaster County, Pennsylvania

spawned the more direct Lancaster Turnpike, a marvel of "modern" engineering and the first turnpike in the country.[35]

Warwick Township, situated in the north of the county just north of Lancaster, exhibited one of the greatest concentrations of Pennsylvania German settlement in all of Lancaster County in 1798 (Figure 2.1). While the earliest settlers here had been both German and Swiss, the borough of Lititz, founded by Moravians in the 1750s in the center of Warwick, had a profound effect on the area's cultural and socioeconomic life from an early date, and this Moravian influence remained significant for a full century after the borough's founding.[36] The strong Germanic element, especially visible in the surviving architecture, produced a high degree of ethnic concentration: in 1798 Pennsylvania Germans comprised over 90 percent of the population, with surnames such as Musselman, Brubacher, Eby, and Grub filling the tax rolls.[37] In Warwick, as in much of Lancaster County, agriculture dominated the local landscape and economy, but milling, including both merchant and custom grist mills, was prevalent.

Superficially, the Warwick tax list describes a rural landscape peppered with a variety of outbuildings, in addition to houses, barns, stables, and kitchens. What distinguished this area from many other Delaware Valley landscapes, in fact, was this very abundance of different kinds of outbuildings with specifically designated functions. These ranged from hay houses, washhouses, distilleries, and storehouses to oil mills and potash complexes. Log construction predominated for most of these buildings in 1798. Log buildings stood on most Warwick farms, where the average property owner maintained a small log house, a large log or combination stone and log barn, and sometimes a log stable. Individuals such as George Hoog, assessed for a small, one-story log dwelling, a barn built of round logs, and a log stable, owned this most typical constellation of farm buildings. Additionally, artisan shops were strewn about many farmscapes, underscoring the importance of supplemental or off-season trades to the mostly agricultural economy. Nearly a fifth of all Warwick property owners were taxed for some type of artisan shop in addition to their dwelling and agricultural buildings. These included currier, joiner, and hatter shops; wheelwright, nail maker, gunsmith, tinsmith, carpenter, and shoemaker shops; and, most frequently, weaving and smith shops. While most of these artisan shops were small log buildings measuring around 300 square feet, such as Lorenz Herchelrot's 20-by-14-foot cooper shop, some, such as Peter Lehnert's and Mathew Kamerer's smith shops, were made either of stone or a combination of log and stone. Other functionally specific outbuildings also stood on many farms. Stone or log springhouses occurred almost as commonly as artisan shops, suggesting the importance of dairying throughout the area. These small, mostly one-story roofed buildings were constructed atop running springs and were used for keeping perishable foods such as milk and cream cool in the summer months and above freezing temperature during the winter. Michael Bohm's 15-by-14-foot log springhouse and George Hollinger's 10-by-8-foot springhouse made of stone typified this building form. Other types of outbuildings, such as detached kitchens, second dwellings, and hemp mills also appeared, but with decreasing frequency.

Although the township of Warwick was overwhelmingly Germanic, some intriguing differences do begin to emerge when we compare the Pennsylvania German majority to the few individuals who were *not* obviously of Germanic extraction. (See Table 1 in the Appendix.) For example, non-Germans in Warwick farmed considerably smaller plots. Their land was also much less highly valued; in fact, their land was usually assessed at around half the total property value of their Pennsylvania German neighbors. While this difference may be related to factors that have little or nothing to do with ethnicity, it nevertheless supports the observations of so many contemporary correspondents that "the Germans are some of the best farmers in the United States, and they seldom are to be found but where the land is particularly good," and that "when the young men of a [German] family are grown up, they generally endeavor to get a piece of land in the neighbourhood of their relations, and by their industry soon make it valuable."[38] Many such observers repeatedly concurred that Pennsylvania German farmers typically not only purchased the best land, often in close proximity to other Germans, but also improved it through careful husbandry.

While there are marked differences in the value of landholdings among ethnic groups in Warwick, comparable variations in the built environment do not appear. Houses and barns owned by the rest of the population were only slightly smaller than those of Pennsylvania Germans, and, contrary to prevailing stereotypes, non-German preferences in building materials actually mimic the supposed characteristics of their Germanic neighbors. In Warwick, barns owned by non–Pennsylvania Germans were also more likely to be built of stone and their houses of logs. Thus, the picture of this ethnic landscape emerging from the 1798 direct tax records seems clear in terms of land use but far less so in terms of the built environment. Warwick's Pennsylvania German landscape did not exactly fit the stereotype of a "back settlement" where "barns are as large as palaces, while the Owners live in log hutts."[39]

Yet although Warwick was predominantly Germanic, much of the rest of Lancaster County was less so. Nearby Hempfield and Conestoga Townships, each with smaller Germanic populations than Warwick, provide useful points of comparison. Located just southwest of Warwick along the Susquehanna River, Hempfield had also been settled primarily by German and Swiss immigrants, many of whom were Mennonites. One of the original townships established in Lancaster County in 1729, Hempfield was named for the economically significant hemp and flax crops, which continued to be cultivated through the eighteenth and early nineteenth centuries. Like Warwick, Hempfield was primarily agricultural; like Warwick, its rural landscape was laced with so many streams that milling became an important element in the local economy.[40]

At the end of the eighteenth century, while strongly Germanic, Hempfield was slightly more ethnically mixed than Warwick. German surnames listed there amounted to around three-fourths, rather than Warwick's nine-tenths, of the total population when the direct tax was assessed in 1798. But although the area's ethnic mix varied more, Hempfield's built environment resembled Warwick's in many ways. As in Warwick, a traveler passing through the area would likely see log buildings most frequently and encounter various types of outbuildings designed for specific functions, although there tended to be fewer of them in Hempfield than in Warwick. (See Table 2 in the Appendix.) These auxiliary buildings ranged from washhouses, bakehouses, smoke-houses, stables, and wagon houses to oil mills, sawmills, hemp mills, granaries, distilleries, and tannery complexes. Only around half of Hempfield's inhabitants owned at least one outbuilding in addition to a small, one-story log house; fewer than half owned a log barn; and just over one quarter owned a stable. Christian String, who lived in a 26-by-32-foot one-story log house and maintained only one other farm building, a 28-by-22-foot log barn, typified many Hempfield residents. The occasional stone or log distillery, spring-house, or kitchen was less common than in Warwick, and only a few people were assessed for such improvements. Similarly, artisan shops such as smith, cooper, and leather currying shops also appeared infrequently. As in Warwick, in Hempfield the greatest differences between Germans and non-Germans again appeared in terms of land quality and quantity. Also, Pennsylvania German houses and barns in Hempfield matched the prevailing stereotype a little more closely than those in more strongly Germanic Warwick: in Hempfield, more Pennsylvania Germans lived in small log dwellings and maintained stone or log barns of impressive dimensions.

Farther south along the Susquehanna River, the built landscape began to change. Conestoga Township, named for the Conestoga River and located south of Lancaster Borough, was the most ethnically mixed township of the three, settled mostly by Germans and Swiss intermixed with a small group of English and Scots-Irish. Here the land and soil quality was somewhat variable. As in Hempfield and Warwick, agriculture and milling were of greatest economic importance, although an iron manufacturing industry would develop in this area in the early nineteenth century.[41]

When the 1798 direct tax was assessed in Conestoga, inhabitants with German surnames comprised an even smaller portion of the total population than Hempfield, amounting to only around two-thirds of the total. Conestogans generally lived in slightly larger log houses and maintained a different constellation of farm outbuildings than their Warwick and Hempfield neigh-

FIG. 2.2. George Stoner's barn on Pequai Creek, Buckhalter's Ferry, Susquehannah River. Barns built of mixed materials and with projecting forebays became more common in the early nineteenth century. Watercolor by Benjamin Henry Latrobe, 1801; courtesy Maryland Historical Society, Baltimore.

bors. (See Table 2.3 in Appendix.) In addition to their houses, most inhabitants owned at least two outbuildings, which typically included a barn; but many were also assessed for a second dwelling or tenant house. Weaving, coopering, wagonmaking, joinery, and smith shops, as well as stables, kitchens, distilleries, and cider houses, occasionally appeared on farmscapes. Like their neighbors to the west and north, Conestoga farmers favored commodious barns, but construction materials for these barns differed from those in Warwick and Hempfield. Substantial barns built of mixed materials, like the one pictured in Figure 2.2, many with stone lower levels topped with log upper stories, appeared much more often.

Material differences between ethnic populations in Conestoga appear to be less than elsewhere in Lancaster County. While the Pennsylvania Germans were consistently wealthier and owned more land than everyone else, the size of the gap is not as great as elsewhere. And although Pennsylvania Germans tended to inhabit larger dwellings than others, the barns owned by both groups were of similar size. Preferred building materials exhibited minimal distinctions between populations.

**Comparison of Total Acreage**

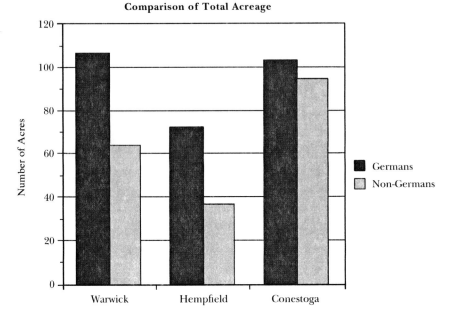

**Comparison of Property Values**

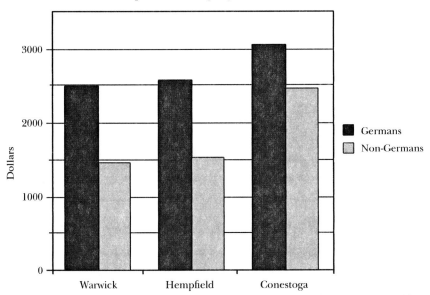

FIG. 2.3. Pennsylvania Germans in Warwick, Hempfield, and Conestoga Townships consistently owned larger plots of more valuable land than did non-Germans.

Several key points thus emerge from the tax lists. First, distinctions are clearest in terms of land ownership and value (see Figure 2.3). The cultural importance of land to the Pennsylvania Germans, suggested repeatedly by contemporary remarks about their propensity to acquire it, is supported by the statistical record. This distinction was most apparent in the more emphatically Germanic townships of Warwick and Hempfield but began to fade in the more ethnically diverse population of Conestoga. Second, Pennsylvania Germans were consistently wealthier than their non-German neighbors. This distinction was clear in all three areas. But what is more suggestive is that, once again, differences between populations are clearest in the most ethnically homogeneous townships and begin to fade as diversity increases. Third, and perhaps most interestingly, ethnic distinctions in the built environment are decidedly less obvious than contemporary accounts imply. Stone was the building material most often associated with Pennsylvania German barns in the historical literature, yet statistically, a Pennsylvania German preference for stone is not uniformly apparent, stone being prevalent in barns and houses only in Hempfield. The notion that Pennsylvania German barns, so often compared to palaces or cathedrals in contemporary accounts, were significantly larger than barns of non-Germans is also not supported by the statistical record.[42] Pennsylvania German barns listed on the 1798 tax records were often, but not always, larger. In fact, the discrepancies are slight in Warwick, the most ethnically homogeneous township; they are most significant in Hempfield; and the stereotype is completely inverted in Conestoga, the most mixed township of the three.

✤ ✤ ✤

A substantially different population mix was found in Sadsbury Township, located in Chester County, to the east along the Octorara Creek, which divides Chester and Lancaster Counties. Sadsbury illustrates a landscape that was far more "English" in the early decades of the republic than Warwick, Hempfield, or Conestoga. It was populated primarily by property owners of English, Welsh, and Scots-Irish extraction at the close of the eighteenth century; almost no Pennsylvania German surnames appear on Sadsbury's tax rolls, and several wealthy and established Quaker families had already made their mark on the agricultural landscape.[43] Like much of Lancaster County, Sadsbury's built environment reflected an agricultural orientation, with barns, springhouses, mills, agricultural and domestic outbuildings, and small artisan shops predominating throughout the region. But once again, when we attempt to side-step the "English eyes" of contemporary observers by comparing Sadsbury to the more Germanic areas of Lancaster County, we find that some of the traditionally held distinctions between ethnic populations fit but others decidedly do not. (See Table 2.4 in Appendix.) For example, the typical house in this "English" landscape, unlike its counterpart in Warwick, Hempfield, and Conestoga, was overwhelmingly likely to be constructed of stone and the typical barn of logs, a choice of building materials that corroborates the observations of so many contemporary correspondents (Figure 2.4). But although most people in Sadsbury owned less-valuable property than most Pennsylvania Germans farther west, the size of the average landholding in this predominantly English landscape significantly eclipsed that of any of the more Germanic townships. Similarly, although the barns in Sadsbury were much smaller, thus bolstering the stereotype of the oversized Pennsylvania German barn, the average house in this more English township was in fact smaller than the reputedly humble houses of the predominantly Pennsylvania German areas.

Because contemporary sources describing "English" landscapes in detail are much more difficult to find, information from the 1798 tax records is especially valuable.[44] But, even though we can compensate somewhat for the "English eyes" of so many contemporary observers, the portrait of the built environment painted by the tax data is just as incomplete for Sadsbury as it is for the more Germanic townships and raises as many questions as it answers. How visible was the Pennsylvania German imprint on the early national landscape? What *was* the role of ethnicity in shaping the built environment, and to what extent was the landscape molded by other complex and interconnected factors such as economy, geography, and priority of settlement? More to the point, to what extent did the houses and barns and fields that contemporary correspondents and tax assessors described actually serve as cultural indicators or badges of ethnic identity?

"Culture," "identity," and "cultural change" are slippery terms to begin with, but all are made more difficult to assess by applying them to an isolated historical moment. While many of the contemporary accounts span a period of years and are the sum of many impressions, the 1798 direct tax assessment by itself, despite its comprehensiveness, is still only a documentary window, consisting of data gathered at a single point in time. It provides only a limited snapshot of a landscape that was undoubtedly in a constant state of flux. But

FIG. 2.4. John Truman house, Sadsbury Township, Chester County, Pennsylvania, photographed in 1986. The Truman house, built in 1789 for a well-to-do Quaker farmer, was larger than most other dwellings in Sadsbury.

because a second direct tax was assessed in 1815 and records of it survive for all of Lancaster County, we can capture not only differences across space but changes in the landscape over time. By comparing the information from both tax assessments for Warwick Township, we can explore the textures of cultural change in one community in greater detail.

The time that had elapsed between 1798 and 1815 spanned nearly a generation. By 1815, Warwick's physical and material landscape had changed. Heightened economic prosperity, largely due to rising wheat prices after the Revolution, was reflected in changing household consumption patterns and in architecture. Just as Cazenove had intimated, though, Pennsylvania German consumption of material goods had shifted only slowly and gradually. As we have seen, in the eighteenth century even the most prosperous Pennsylvania German farmers and tradesmen inhabited only sparsely furnished houses. They sank most of their wealth into land, livestock, bonds, and notes. As so many contemporary observers had remarked, the typical German farmer or artisan filled his house with kitchen tools, farm implements, grain, and food storage containers but invested little in the furniture or decorative household goods that Cazenove deemed necessary for a "civilized" lifestyle. Only the most Anglicized and prosperous Pennsylvania German farmers had owned houses filled with furniture and "English" goods. Although economic prosperity grew, domestic consumption among Pennsylvania Germans remained at a relatively low level even into the second decade of the nineteenth century, increasing only gradually as regional prosperity accelerated. In his study of Berks County and Lancaster County household inventories, Scott Swank found that a common constellation of several fairly expensive items did appear in numerous Pennsylvania German inventories and remained traditional Germanic symbols. This constellation of goods consisted of clocks, beds and bedsteads, clothespresses, kitchen cupboards, and stoves—goods which were typical of most middling as well as wealthy Pennsylvania German households in the eighteenth and early nineteenth centuries. Acculturation, as evidenced by increasing mention of more "English" goods, such as tea tables, silver, maps, sets of chairs, and tea equipage, proceeded slowly and unevenly.[45]

Acculturation effected changes to the built environment as well. The most significant change to the Warwick landscape by 1815 appeared in the distribution of the land itself. (See Table 5 in Appendix.) As the population increased and as succeeding generations inherited partitioned land, the average farm size declined markedly for everyone: between 1798 and 1815, the size of the average land parcel decreased by nearly one-quarter. At the same time, though, houses and barns expanded somewhat in size. Although a farmer was still likely to own the same constellation of buildings as he had in 1798—a house, a barn, and perhaps another outbuilding, such as a stable, artisan shop, or tenant house—this period saw the appearance of multifunction industrial buildings, such as John Keller's combination snuff and carding mill. Some of these added living space in or adjacent to a work building. Barbara Sheaffer's wooden grist mill was furnished with an adjoining dwelling house, and John Yunt's two-story stone combination building included a "gristmill, oil mill, and dwelling house under one roof." An assortment of domestic outbuildings such as cornhouses, stillhouses, and smokehouses proliferated, but the relationship between these outbuildings and the functions they housed began to change. Multifunction outbuildings—like Peter Becker's consolidated bakehouse-washhouse, Peter Buch's combination stable and chair house (carriage house), and Jacob Eberly's combination cornhouse and pigsty—began to multiply. The heirs of Jacob Erb owned a stone washhouse with a springhouse below; John Bear was assessed for an outbuilding described as a "corn crib and wagon shed in one crib"; John Bender maintained a combination wooden springhouse and toolhouse; and John Sprinkler owned a combination wooden wash and bake house. Although many such buildings brought together several domestic and agricultural functions, they also occasionally consolidated living space and rougher work functions under one roof (Figure 2.5). Several dwellings in Warwick were combined with springhouses and washhouses, such as John Metz's one-story stone dwelling "with wash and springhouse below," and Andrew Grove's wooden tenement "having a spring and wash house below." In addition, these ancillary living spaces were occasionally occupied by tenants, often of little means. Such was the case of Samuel Showmaker, who was assessed for a one-story wooden washhouse with a one-story wooden dwelling house adjoining, "occupied by a poor tenant who pays no rent." Likewise, Henry Geppel owned an 18-foot-square wooden tenement that was "inhabited by 3 old maids rent free." While a few such multifunction outbuildings had been present in 1798, they had clearly become more common by 1815.[46]

FIG. 2.5. Outbuilding, Conestoga Township. Some outbuildings combined multiple functions under one roof. The lower story of this building accommodated domestic and agricultural work such as rendering and dairying, but the upper story, neatly finished and provided with light, air, and heat, would have served an artisan function like weaving or joinery. Photograph by Bernard L. Herman.

Building materials had also changed. Although log and timber dwellings were still the most typical, other construction materials, including frame and brick, had grown in popularity. Buildings such as Jacob Graber's 32-by-30-foot two-story brick dwelling and Benjamin Stehman's 16-by-25-foot two-story frame house were becoming much more common. Still, traditional construction materials associated with Lancaster County's Germanic element, like clay tile and thatch for roofing, persisted in Warwick even into the second decade of the nineteenth century. James Conrad was assessed for "one smith shop of timber covered with tiles;" similarly, assessors recorded Jacob Spohnhauer's "weavers shop of timber fifteen feet by twelve one story thatched with straw."[47]

Barns had changed more visibly. Log barns still proliferated, but by 1815 Warwick farmers increasingly favored somewhat larger barns built entirely of stone, frame, brick, or a combination of these materials. Buildings such as Peter Holl's 40-by-22-foot barn with stone lower story and wooden upper story and Jacob Hohstetter's 100-by-30-foot combination stone and wood barn were becoming increasingly common. Also, barns with projecting forebays or "overshots" were built with increasing frequency. These oversized overshot barns, such as John Pfautz's "new barn one hundred and six feet by forty feet in depth, overshot, lower story stone upper Brick" were becoming widespread enough that they began to be distinguished from earlier models, as in William Shutt's 40-by-22-foot wooden barn, which the tax assessor described as "old fashion."[48]

Warwick's Pennsylvania German population also seems to have been experiencing cultural change, at least based on the number of Germanic surnames on the rolls. Just as Isaac Weld had predicted in 1800, a degree of acculturation had begun to occur, and the number of Germanic surnames was clearly on the wane. No matter whether the catalyst for such change was exposure to other cultures, the passage of time, gradual generational transitions, or overall population movement, the number of Germanic surnames decreased in the first decades of the nineteenth century, and by 1815, Warwick's Pennsylvania German population had declined from over 90 percent of the total population to 81 percent.[49] Equally significantly, a number of measurable gaps between Pennsylvania German and other farmers had widened appreciably, throwing some ethnic distinctions into much sharper relief while clouding others further. (See Table 6 in Appendix.) Compared to their neighbors, by 1815 Pennsylvania Germans in Warwick had grown much richer, had built vastly larger barns, and, contrary to the prevailing stereotype, were living in significantly larger timber houses. In addition, their houses were much more likely to be built of stone or brick than those of their neighbors.

The greatest measurable distinction between Germans and non-Germans was barn size; although larger barns had become popular throughout the entire Warwick population by 1815, Pennsylvania German farmers there were building barns that were dramatically larger than those of their neighbors. Pennsylvania's barns had long impressed many observers with their size. Just as noticeably, these barns were very well organized buildings where everything had its designated place. "In the middle is the threshing-floor," wrote Thomas Anburey in 1789, "and above it a loft for the corn unthreshed; on one side is a stable, and on the other a cow house, and the small cattle have their particular stables and styes; and at the gable end of this building there are great gates, so that a horse and cart can go strait through: thus is the threshing-floor, stable, hay-loft, cowhouse, coach-house, &c. all under one roof."[50] The larger barns that began to proliferate in the early nineteenth century were bank barns: large, two- or three-story multipurpose barns built into earthen embankments and designed to consolidate as many farm functions as possible into one large, efficient building. Bank barns, which first began to appear in Pennsylvania late in the eighteenth century and grew common by the early 1800s, spread at least 30 to 40 feet wide, "for giving room to turn waggons within the house" according to agricultural reformer John Beale Bordley.[51] William Cobbett, writing in 1818, waxed enthusiastic about Pennsylvania's bank barns, calling them "very fine buildings." In the Pennsylvania backcountry, Cobbett found "barns of *stone,* a *hundred feet* long and *forty wide,* with two floors, and raised roads to go into them, so that the waggons go into the *first floor up-stairs.* Below are stables, stalls, pens, and all sorts of conveniences."[52]

In 1815 in Warwick, the overwhelming majority of these impressive barns were owned by Pennsylvania German farmers, and these agricultural monuments had become closely associated with the Pennsylvania German population. Ninety-five percent of all barns that were at least 30 feet wide by 60 feet long belonged to Pennsylvania German farmers that year. This pattern extended beyond Warwick. All of the barns in the predominantly Germanic townships of Lancaster County were statistically more than half again as large as those in the mostly English townships. More suggestively, it appears that, for Pennsylvania Germans in Warwick, at least, a marked shift in longstanding cultural values was beginning to occur: the average barn size had increased even as the average size of their land holdings decreased. While good land had always been important to Pennsylvania Germans, the oversized barn now seemed to be of equal, if not greater, importance. Other Germanic townships throughout the county tended to follow the same pattern of smaller average land parcels but correspondingly greater total property values.[53] Although this same inversion was also prevalent throughout the general population and can easily be linked to the rise of the bank barn in early nineteenth-century agricultural literature, it was far more apparent among Pennsylvania German property owners. Had the oversized bank barn become, for Pennsylvania Germans, a new and transformed boundary or cultural marker—a new and potent embodiment of "Pennsylvania Germanness" that had eclipsed the symbolic place of land? Were the Pennsylvania Germans, in response to widespread pressure to assimilate, "dropping their heaviest cultural anchors" and inventing themselves and their ethnicity anew with these distinctive barns?[54] Or was this a different kind of symbol, one not necessarily related to ethnicity but rather to prosperity or geography or priority of settlement—characteristics that may or may not be related to ethnicity? Once again, although the tax records provide a great deal of suggestive evidence, they do not reveal the clear ethnic distinctions that eighteenth-century observers so often noted.

Despite persistent comments regarding the distinctiveness of the Pennsylvania German built environment then, ethnic differences are not as apparent in the statistical record of the 1798 and 1815 taxes as the contemporary accounts might lead us to believe. *Perceptions* of difference apparently eclipsed the architectural substance of the countryside. But why? As a partial answer to this question, James Lemon, noting that stereotypes of Pennsylvania German farming practices were firmly entrenched by the eighteenth century—and have endured—suggested that early Pennsylvanians inherited a particular set of attitudes from England by which they judged others. Late-eighteenth-century writers also differentiated many national cultural traits rather sharply because they tended to transfer impressions from one area to the whole, effectively exaggerating them. For instance, because the Lancaster Plain was regarded as Pennsylvania's most productive agricultural region, and because German Mennonites inhabited that area, *all* Pennsylvania Germans came to be linked with agricultural excellence. Ethnic stereotypes were also perpetuated by the widespread practice of "borrowing" descriptions from existing publications; many eighteenth-century travel writers copied shamelessly from one another, often without crediting their sources. Finally, contemporary correspondents may have fostered notions of German distinctiveness for political or philosophical reasons. Tracts praising Germans, such as Benjamin Rush's famous commentary, may have been written to gain German support for American nationalism and the Constitution. Existing stereotypes also may have become more deeply embedded because Pennsylvania Germans best personified the prevailing Virgilian ideal of the noble and virtuous farmer.[55]

The information from the tax records is suggestive but still leaves the question of assimilation largely unresolved. How distinctive *were* the Pennsylvania Germans by 1798, and how much acculturation or creolization had already occurred? By the time the 1798 direct tax was assessed, people of Germanic extraction had farmed Pennsylvania soil for over a century. Several generations of Pennsylvania German farmers had lived, died, and made their own imprint on the landscape. And Lancaster County Germans, although still retaining numeric significance, were already beginning to be absorbed into a broader Anglo-American regional culture, fraught with its own nascent iconography.

In his study of the German Massanutten settlement in the Shenandoah Valley of Virginia during the same period, Edward Chappell argued that although many ethnic characteristics survived acculturation, Germans there began to reject some of the most visible symbols of their background shortly after 1800 by abandoning their traditional Rhenish house form and discarding much of their standard German language. But although they rejected or altered the most conspicuous symbols of their culture, other culturally rich traditions, such as decorative gravestone motifs, woodwork in domestic interiors, *Fraktur,* painted barn decorations, and regional Germanic dialects not only survived but flourished.[56] Similarly, Kathleen Conzen and others have noted how ethnic symbols and rituals acquire new, different, and sometimes more culturally charged meanings during times of societal crisis.[57] In exploring the intertwined notions of cultural persistence, assimilation, and change, Chappell and others have recognized that cultures invest different types of objects with varying levels of meaning at different points in their accommodation and that they may place more meaning in certain objects at times when their existence seems most precarious. In other words, as the pressure to assimilate increases, ordinary material objects such as houses, barns, and fields may take on heightened significance—or different meanings entirely—and may begin to serve as important cultural repositories or boundaries.[58] Or, put another way, the Massanutten Germans became the most "German" just before they began to be absorbed into the mainstream culture in earnest.

Between 1750 and 1840, the boundaries of Pennsylvania's German culture were gradually shifting. Pennsylvania Germans during those years formed, according to one historian, a hybrid culture—a community that was "both European and American, both English and German—a fragment caught between two worlds."[59] At the close of the eighteenth century, when the 1798 tax was assessed, Pennsylvania's Germans were also in the process of renegotiating the borders of their community and mapping their cultural boundaries anew, often with the ordinary objects of everyday life. Yet even if these boundaries were undergoing redefinition, why are the differences that are so prevalent in the contemporary accounts not more apparent in the built environment of 1798? What *was* the basis of the reaction that contemporary correspondents experienced in the Pennsylvania German countryside? Are we dealing with tangible material objects—with houses, barns, and fields—or are we dealing with something else—the much slipperier notion of *perceptions* of objects?

While houses and material objects may have seemed to be transparent badges of identity to some contemporary observers, the degree to which they signified the same meanings for their occupants is less obvious. To what extent was ethnicity or cultural identity actually expressed in architecture or in other material genres, and to what extent can we capture those meanings— or even uncover those distinctions—from surviving evidence?[60]

Although surviving architecture seldom presents a completely balanced view of historic landscapes, a significant number of still-extant eighteenth-and early-nineteenth-century buildings in Warwick township provoke us to ask similar questions, but this time from a different body of evidence. In Warwick, as elsewhere, the majority of surviving eighteenth-century buildings represent the lifestyle of economic elites—typically the wealthiest slice of the population—rather than the middling or lower sorts. Extant dwellings belonging to poorer or moderately well-to-do individuals survive in much smaller numbers. As the tax lists indicate, Warwick's late-eighteenth-century architectural landscape was fairly complex, and a wide range of housing options was available; this is evidenced also by several extant dwellings that span nearly half a century. While all of the dwellings discussed below were built and occupied by individuals with Germanic surnames and affiliations and many incorporate architectural features often linked with Germanic building traditions, most represent architectural solutions that are complex and difficult to categorize. Warwick's surviving buildings evince both Germanic and non-Germanic architectural features in varying combinations, but they also suggest, in ways that the tax lists and contemporary accounts do not, something of the complexity and unevenness of the acculturative process.

Even though its first-period first-floor plan was altered in the twentieth century, the Christian Eby house (see Figures 2.6 and 2.7), built in 1769 along the Hammer Creek, retains clear evidence of the type of classic three-room floor plan most typically associated with German settlement areas. The house, adjacent to a mill site that is reputed to have been in operation as early as the 1740s, is a three-bay, two-and-one-half-story cut limestone house ornamented with round brick ventilators in the attic gables and brick segmental arches over the first floor window openings. A frame wing built onto one gable is a later addition. The present interior is repartitioned in places, but evidence of the original plan is very clear. With its original interior arrangement consisting of a front to back kitchen, or *küche,* furnished with opposing entrances, a squarish parlor, or *stube,* and a smaller rear room, or *kammer,* arranged around a central chimney, the Eby house typifies the classic German *flurküchenhaus,* or three-room plan—most likely the type of dwelling that Schoepf and others so readily labeled a "German" house. In a classic three-room-plan house, the kitchen was heated by a large cooking fireplace that backed up to the parlor. In such dwellings, the front door of the house also opened directly into the kitchen, which contained the hearth. This meant that the primary working and social space in the house was publicly accessible. The parlor had no fireplace but was heated instead with a closed five-plate or jamb stove that was fueled through a small, squarish opening in the back of the kitchen fireplace. While the original kitchen fireplace in the Eby house has been demolished and end chimneys have been added, evidence in the cellar and attic reveals the placement of the original central stack. The heavy roof framing—a truss system consisting of mixed common and principal rafters—and the vaulted, spring-fed stone storage cellar beneath the parlor are other features that are usually associated with Germanic building traditions.[61] The original stair, most likely a winding stair that was located semipublicly in the rear corner of the *küche* opposite the main cooking fireplace, has also been demolished and a

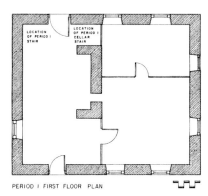

PERIOD I FIRST FLOOR PLAN

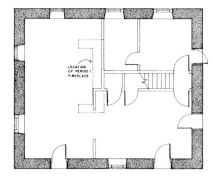

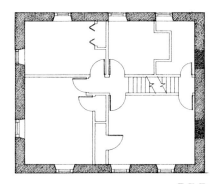

window added in its place on both floors. Access to the second floor is now provided by a more private enclosed straight staircase located in the center of the house, on what was the west wall of the original *stube*. Other evidence of earlier configurations survives. A second-floor door still bears the German inscription "CHRISTIAN EBI 1769," "MAGDALENA EBISIN," and "REBUILD BY JACOB SHEAFFER 1874." Despite the fact that assessors failed to note any dwellings of the same dimensions as the Eby house on the 1798 or 1815 tax lists, all of the architectural evidence—nail technology, plan configuration, and roof construction—corroborates the 1769 construction date painted on the second floor door.

The size of the house and its stone construction as well as the level of original finish also suggest the prosperity of the original owners. Large stone dwellings such as the Eby house were clearly linked with elites and are the most likely to have survived.[62] Dwellings that belonged to Pennsylvanians of more middling circumstances are much rarer, but one log example illustrates the way a more modest dwelling with a similar three-room floor plan was originally built and subsequently altered. The log and frame Rudy house represents a type of dwelling that was once common. In 1815, Charles Rudy's taxable wealth placed him squarely in the middle third of the population. The original log section of the roughly 26-by-30-foot dwelling, which dates to the mid-eighteenth century, consisted of a classic three-room *flurküchenhaus* plan, much like that of the Eby house, arranged around a central chimney (Figure 2.8). The original vaulted storage cellar still survives beneath the old *stube* and *kammer.* Some time around the first decade of the nineteenth century, the log end wall of the original *küche* was demolished and the house was extended with a frame addition, so that now it is difficult to see the original spatial arrangement as clearly from the exterior. When the original chimney was demolished, a corner fireplace was built in the old *kammer,* and a gable-end fireplace furnished with a bake oven was built in the new kitchen, in the frame section. Fi-

FACING PAGE:
FIG. 2.6. Christian Eby house, Warwick, photographed in 1992. This dwelling originally had a central chimney and a classic three-room floor plan (see Fig. 2.7).

FIG. 2.7. First period first-floor plan (*top*) and existing first- and second-floor plans (*bottom left and right*) of Christian Eby house, Warwick. When the central chimney was removed, the original three-room plan was reworked.

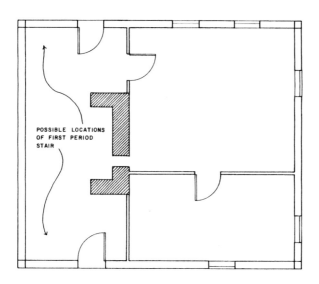

POSSIBLE LOCATIONS
OF FIRST PERIOD
STAIR

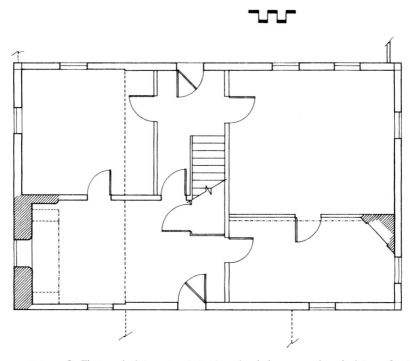

FIG. 2.8. First period (ca. 1750) (*top*) and existing second period (ca. 1805)
(*bottom*) first-floor plans of Rudy house, Warwick. The original plan of this log
dwelling consisted of three rooms arranged around a central fireplace.
The house was remodeled in the early nineteenth century.

nally, a new stair and entry were partitioned off from the old *küche*. Although the remodeling of the Rudy house clearly created a more "Georgianized" plan, the interior finish that dates to this early-nineteenth-century remodeling includes several strongly Germanic features, such as a red-painted *küche* embellished with white scratch-painted stars and a decorative splatted stair balustrade.

Built just after the mid-eighteenth-century in-migration of new German settlers to Pennsylvania had peaked and had started to decline, the Eby and Rudy houses still exemplify some of the most classic features customarily associated with Pennsylvania German architecture: the three-room *flurküchenhaus* plan with a central chimney, provision for food storage within the house, heavy roof framing, and Germanic decorative features. But although central-chimney three-room-plan houses like these took root and flourished in Pennsylvania, they were not particularly widespread in Germany, and they were certainly not the only house forms utilized in the German-settled areas of Pennsylvania.[63] Traditional German houses with two- and four-room plans were widely built as well.

In dwellings such as the Eby and Rudy houses, several Germanic features would have been apparent from the outside and would also have been visible in both public and private interior spaces. But toward the end of the eighteenth century, elite Pennsylvania German houses that combined traditional "German" features with elements more commonly associated with "English" Georgian, architecture began to appear with increasing frequency.[64] Four dwellings built for prosperous German farmers between 1784 and 1813 illustrate the complicated process of negotiation between Germanic and Georgian architectural features and suggest that these now-intermingled architectural traditions had begun to take on new and complex forms.

The Peter Kreiter house, a two-story dressed-limestone dwelling near the center of the Moravian borough of Lititz, was built in 1784, as documented by a relief-carved datestone on the second-floor facade. With its off-center chimney, three-bay facade, stone segmental arches, and steep roof pitch suggestive of heavy trussed roof framing within, the Kreiter house presents a strongly Germanic appearance from the exterior (Figure 2.9). And like the majority of Warwick's other surviving eighteenth-century dwellings, the Peter Kreiter house also represents an elite form.

Despite its exterior appearance, however, the interior reveals an intermingling of Germanic and Georgian features. Measuring 28 feet by 32 feet, the

FIG. 2.9. Peter Kreiter house, photographed in 1992. The house, built in 1784, stands near the center of the Moravian town of Lititz.

house presents its narrower dimension to the front. The interior is arranged, not as a variation of the three-room *flurküchenhaus* plan, but instead in a classic Georgian side-passage plan, with a narrow hall running the full depth of the house and two equal-sized, squarish rooms placed front to back beside the passage (Figure 2.10). Instead of a winding stair tucked into the rear of the *küche*, a straight stair is located prominently near the rear of the passage. Still, the stair balusters consist of the flat, vasiform splats that are customarily associated with regional Germanic building traditions rather than the turned

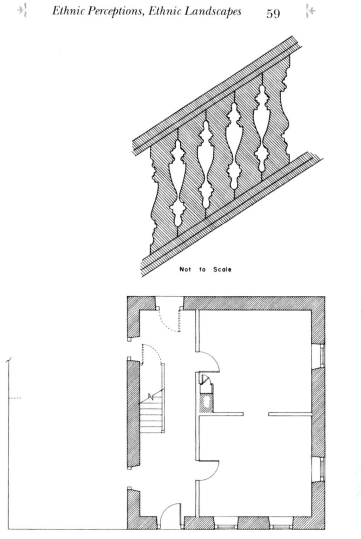

Not to Scale

FIG. 2.10. Peter Kreiter house, Lititz. The interior reveals a combination of
Germanic features and Georgian ones.

balusters that are more typically associated with Georgian forms. Evidence of an original fireplace, now a closed stack, survives in the corner of the rear room. Although the partition between the two first-floor rooms was clearly reworked in the early twentieth century and that which separates the passage from these two rooms may also have been altered, the placement of these partitions, and thus the current plan configuration, judging from extant cellar bearing walls, appears to be original. Furthermore, double-sided relieving arches in the cellar confirm that a fireplace was either once situated near the middle of the passage or was originally intended but was never actually built there. The narrowness of this passage—it measures only 8 feet wide—makes the placement of a fireplace opposite the stair both unwieldy and unlikely; nevertheless, the stair itself appears original, although perhaps not in its original location. Still, even if the stair, clearly one of the most expressive architectural elements surviving from the late eighteenth century, was moved from elsewhere in the house and reused, its prominent "new" placement at the center of the entry passage signals a blatant commingling of Anglo-Georgian and Pennsylvania German features. Similarly, several expressive Pennsylvania German–style door latches survive in fairly prominent locations throughout the house. Finally, and perhaps most revealingly, the architectural evidence suggests the presence of both a formal English parlor, heated with a fireplace, *and* a German *stube*, heated with a stove.

In elite dwellings like the Kreiter house, Pennsylvania German and Anglo-Georgian elements were appropriated and recombined in various ways. Instead of being eliminated or overshadowed, stylistic features as well as some aspects of plan configuration were often reworked into new, German-Georgian forms. Several other elite German-Georgian dwellings built during the same thirty-year period, all with similar floor plans, illustrate a similar process of negotiation.

The Christian and Barbara Reist house was built in 1795 for a prosperous Mennonite farmer and his wife: on the 1798 tax, Christian Reist was listed among the wealthiest 10 percent of property owners in Warwick. In addition to their new two-story stone dwelling, the Reists were assessed for a second dwelling, an 18-by-12-foot stone kitchen, and a 30-by-30-foot combination stone and log barn.[65] By the time of the 1815 assessment, Christian Reist had passed on, but the house was listed in the name of his widow, who was also assessed for an additional one-story "timber" house, a large stone overshot barn, an adjoining one-story stone kitchen, and a log stable. From the exterior, with its external symmetry and end chimneys, the house exhibits the balanced five-bay, two-story facade that Johann Schoepf would readily have linked with English Georgian architectural traditions. Paired German-inscribed datestones set in the facade at the second-story level provide the only really blatantly Germanic exterior feature. The interior spaces exhibit a blending of Georgian and Germanic elements (Figure 2.11). The central hall, symmetrical four-room floor plan, and corner fireplaces with gouge-carved crosseted mantels all nod toward elite Georgian architectural models rather than traditional Pennsylvania German forms. Still, the house incorporates several Germanic features, including elaborate tulip-form iron door latches, a vaulted food-storage cellar below the kitchen addition, a first-floor *stube* and a *kammer*-like room, both of which always lacked fireplaces and were apparently heated with stoves, and cellar ceilings insulated with packed clay and secured with wide wooden boards beneath the entire main block.[66]

Similarly, the Philip and Barbara Friedrich house, built in 1797 for a well-to-do miller and his wife, exhibits few overtly Germanic features on the exterior except for the carved German datestone on the west gable. Like the Reist house, the Friedrich house has a central hall, symmetrically spaced openings, and end chimneys "after the English plan." Now a two-story, five-bay, gable-roofed brick dwelling with a 23-by-32-foot brick rear addition (Figures 2.12 and 2.13), the Friedrich house was listed on the 1798 tax records as a 40-by-34-foot brick house with a 20-by-20-foot stone kitchen, a large stone barn, a stone grist mill, and a sawmill. The Friedrichs sold the house soon after construction; by 1815 it was owned by John Keller and the assessment showed the house, barn, mills, a tenant house, and a stone kitchen of slightly different dimensions, listed as "one stone kitchen joining first house one-story 15 feet square." As with the Reist house, the adjoining kitchen enumerated on the 1815 tax assessment apparently stood near the same spot that the rear addition now occupies. The Friedrich house also illustrates the stylistic transition between the Georgian and Federal periods, including a four-room center hall floor plan, a balanced five-bay facade, molded crosseted door surrounds, an open stairwell with gouge-carved arch, an elaborate gouge-carved mantel, and gouge-carved window sill pendants (Figure 2.14). Yet, as in the Reist house, two of the four main rooms were evidently heated with stoves rather than fireplaces. Significantly, both of these stove rooms were rear rooms. In addition, the least public parts of the house exhibit several interior features commonly associated with Germanic building traditions, including a vaulted food stor-

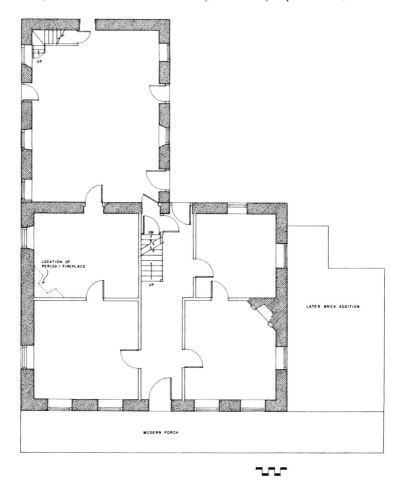

LOCATION OF
PERIOD I FIREPLACE

LATER BRICK ADDITION

MODERN PORCH

FIG. 2.11. Christian and Barbara Reist house, Warwick. First-floor plan, main block. In the Reist house, as in other elite Pennsylvania-German houses, Germanic and Georgian features were combined.

age cellar under the present rear addition—possibly the very cellar that stood under the original stone kitchen listed on the 1798 and 1815 tax lists—and a paled cellar ceiling under the 1797 portion of the house.[67]

A short walk from the Friedrich house is the John and Elizabeth Pfautz house, built sixteen years later and possessing a similar blend of traditional Germanic and English elements. A visitor approaching this house would observe a symmetrical, two-story masonry dwelling with a central hall and "English-style" end chimneys. Like Philip Friedrich and Christian Reist, John Pfautz was one of the wealthiest property owners in Warwick; in 1815, he was

FIG. 2.12. Philip and Barbara Friedrich house, Warwick, photographed in 1992. Built in 1797, the Friedrich house exhibits a symmetrical facade and central hall plan but also includes several Germanic features, such as a vaulted cellar and two first-floor rooms that were evidently heated with stoves instead of fireplaces.

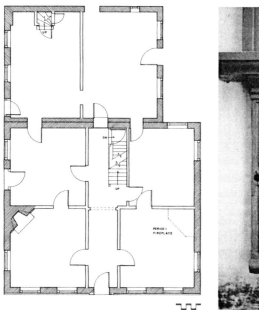

FIG. 2.13. Philip and Barbara Friedrich house, Warwick, first-floor plan. The two front rooms were heated with corner fireplaces, but stoves probably heated the two rear rooms.

FIG. 2.14. Philip and Barbara Friedrich house, Warwick. Detail of gouge-carved window sill pendant.

assessed for his new two-story brick house (Figure 2.15A) in addition to a large two-story stone house, two barns, an adjoining one-story stone kitchen, a stone springhouse, two wood tenement houses, and a large wooden haymow. As with the Reist and Friedrich houses, the only really obvious Germanic exterior features on the Pfautz house are a set of paired German datestones on the second-floor level of the front elevation (Figure 2.15B). Otherwise, the symmetrical five-bay brick facade, Flemish bond brickwork, end chimneys, beaded and gouge-carved front entry with fanlight, center hall floor plan, interior moldings, and gouge-carved cornice speak more strongly to the prevailing Federal style than they do to earlier Germanic building traditions. The interior is divided into four rooms on either side of a wide center hall (Figure 2.16), and while corner fireplaces have been added to the two front rooms, evidence in the cellar indicates that the two rear rooms, as in the other German-Georgian houses, were originally furnished with fireplaces while the two front rooms were apparently heated with stoves. The house now lacks the attached kitchen wing that was enumerated on the 1815 tax record, but an original cooking fireplace with evidence of an external bake oven survives in the rear room to the left of the stair. Although the Pfautz house lacks the elaborate Germanic hardware and vaulted storage cellars that were present in some of the other houses, it retains one very traditional Germanic feature in one of the least public spaces of the house: paled clay-packed insulation survives in much of the cellar ceiling.[68]

While Germanic and Georgian elements are worked out in slightly different ways in the Reist, Friedrich, and Pfautz houses, the floor plans of the structures are similar in many respects, and the dwellings share the characteristics of a balanced symmetrical facade that appears from the exterior to be a full Georgian plan, a center hall, end chimneys, and two rooms without fireplaces on the first floor. Some highly ornamental Germanic features, such as elaborate iron door latches and carved German datestones, remained prominent; still, some of the most conservative German elements, such as the open *küche* and the central chimney, were excised from the most public spaces, while other traditional Germanic features, such as vaulted food storage cellars and paled insulated ceilings, persisted in the least public areas. The same blending of architectural traditions and construction features that is evident in these three houses is also apparent in several Cumberland County, Pennsylvania, dwellings of the same period which exhibit balanced facades and remarkably similar floor plans.[69] Likewise, the Kreiter house, with its side-passage plan, decorative splatted center stair, and squarish, *stube*-like, front parlor with no fireplace, represents a negotiation between Pennsylvania-German and Anglo-Georgian influences.[70]

The persistence of the big, square *stuben* in these "Georgianized" floor plans, even into the second decade of the nineteenth century, is particularly suggestive. The cultural significance of the *stube* has been well documented by William Woys Weaver, who argued that Germans "viewed stoves as an absolute necessity, as material proof of domesticity" and that "a room without a stove was not a *stube*."[71] Or, as Richard Weiss has remarked: *"Ohne Ofen, keine Stube, ohne Stube, keine Hauslichkeit"* (No stove, no *stube*, no *stube*, no home).[72] This appropriation of Georgian features mixed with traditional and sometimes culturally charged Pennsylvania German elements represents, not a diminution or a simple fusion of traditional forms, but, rather, the creation of a new hybrid form, or a kind of architectural creolization. And in architecture, such creolization tended to occur, as Theophile Cazenove had observed over two centuries earlier, slowly, gradually, and from the outside in. While many of these buildings presented balanced "Georgian" exteriors to passersby, their interiors revealed a more complex and subtle struggle to redefine Pennsylvania Germanness. As Henry Glassie has argued, "The skins of houses are shallow things that people are willing to change, but people are most conservative about the spaces they must utilize and in which they must exist. Build the walls of anything, deck them out with anything, but do not change the arrangement of the rooms or their proportions. In those volumes—bounded by the surfaces from which a person's senses rebound to him—his psyche develops; disrupt them and you can disrupt him."[73] These creolized Pennsylvania German dwellings, typically linked with elites, utilized elements that were present in both traditional Pennsylvania German and Anglo-Georgian architecture. They also drew from architectural traditions that had been well-developed in Germany, for the Georgian plan was part of an international Renaissance style that had gained some currency there as well as in Pennsylvania.[74] Like the evidence that can be gathered from the 1798 and 1815 tax lists, the indications of ethnic distinctions revealed in these German-Georgian buildings are, in most cases, subtle and not particularly apparent from the exterior, emerging only from a more sustained examination of the interiors. The evidence from the surviving buildings not only suggests a great deal about cultural persis-

FIG. 2.15A. John and Elizabeth Pfautz house, Warwick, built in 1813, photographed in 1992.

FIG. 2.15B. John and Elizabeth Pfautz house. Detail photograph shows original paired German datestones set in facade at second-floor level.

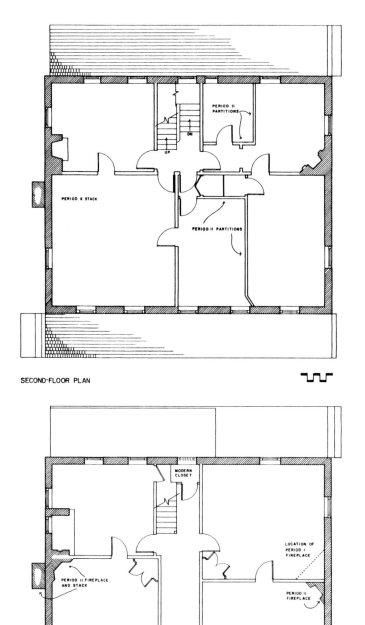

SECOND-FLOOR PLAN

PERIOD II PARTITIONS

PERIOD II STACK

PERIOD II PARTITIONS

UP
DN

FIRST-FLOOR PLAN

MODERN CLOSET

PERIOD II FIREPLACE AND STACK

LOCATION OF PERIOD I FIREPLACE

PERIOD II FIREPLACE

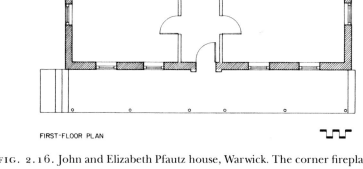

FIG. 2.16. John and Elizabeth Pfautz house, Warwick. The corner fireplaces that appear in the two front rooms on the first-floor plan (*bottom*) are not original. Evidence in the cellar indicates that these rooms were heated with stoves.

tence, negotiation, and the complex process of cultural change, but also supports what the tax lists reveal—that ethnic distinctions in the built environment are much less obvious than contemporary accounts imply.

While James Lemon has suggested that the "English eyes" of many contemporary observers colored their perceptions and influenced their accounts, it may be useful here to return to Dell Upton's notion of texture in the landscape. Upton has emphasized the need to consider not just the artifact but the connection *between* person and artifact, not just the objects that are the traditional focus of material culture studies but the textures that form the interstices of everyday experience. He refers to the smells, sounds, additional sights, and emotional responses which, although they are accidental byproducts of culturally determined behavior, nonetheless often directly affect the way we interact with the physical world. While these are the aspects of our surroundings that have been called "invisible" material culture, they are still features that are culturally determined, and these incidental cultural byproducts intrude on our consciousness, mold our reactions, and require a new range of responses. Our responses are, in turn, interpreted in cultural terms.[75]

Although the responses of Schoepf, Whitelaw, Cazenove, and other contemporary correspondents had a firm grounding in the visual world, they also may have been based on multifoliate textural cues that, however fleeting and ephemeral, still expressed culture, borders, and boundaries. These textures— key elements in the mental mapping process—rarely found their way into the 1798 tax assessments. Although assessors recorded the late-eighteenth-century landscape with astonishing comprehensiveness, the tax lists do not describe architectural detail beyond exterior dimensions, materials, number of stories, and perhaps age. Some of the most dramatic and expressive Pennsylvania German dwellings that survive on the landscape today were recorded in exactly the same way as their more restrained English counterparts. While assessors were usually able to penetrate the most formal and public facades of the properties they encountered in their efforts to record all standing buildings, the terrain they mapped was still fundamentally an exterior rather than an interior landscape. They rarely recorded ornamental details or structural conditions, almost never ventured inside of buildings, and stopped far short of differentiating between German and non-German architectural characteristics. The tax rolls reveal little about the interiors of the dwellings they describe and say nothing about whether their walls and doors were "daubed with red" throughout, or whether their rooms were furnished with German or English goods, filled with potatoes and turnips on the floor, suffused with the aromas of stewed pork, sour milk, and warm slaw, or heated with the "great four-cornered stoves" that Johann Schoepf described.

Despite its comprehensiveness and precision, then, the 1798 tax record is ultimately—and paradoxically—an abstraction of the built world, a text that is seductive in its thoroughness but fundamentally devoid of texture, a landscape narrative that heightens our awareness of objects that have since vanished from the countryside but that, in the end, reduces the sense of place. Surviving buildings, which expose much of the very texture that the direct tax account lacks—interior features such as the big, square *stuben* and decorative splatted stair balusters, for example—in turn reveal a very limited band of historical reality, because they represent elites and usually lack their original contexts. The ethnic landscapes described by Schoepf and his contemporaries amount to a landscape reality colored by complex perceptions of ethnicity; they are caught in the interplay between perceptual measures of difference and the objects of the material world. In our efforts to make cultural sense of the built environment, we need to remember that we are dealing not just with artifacts but with human agency and intention, not just with the built stuff of the material world but with the way people interacted with—and responded to—architecture and objects in the landscape. The historic landscape amounts to much more than a complex artifact assemblage. Landscape, after all, is fundamentally a communicative medium, and, like every communicative system, it is nuanced with layers of performance, dialect, and inflection—the features that distinguish language from speech.

# 3 CHAPTER

# *Landscape on the Margins*

O n a warm April day in 1829, a farmer was plowing the land he rented in northwestern Sussex County, Delaware, just over the Maryland state line. As the plow slowly recaptured land that had once defined the margins of last year's cultivated field, his horse suddenly sank down deep into the dirt. When the beast struggled to regain its footing, the farmer saw an odd flash of blue in the gaping hole below. Digging deeper, he uncovered more blue: a painted blue pine chest lay shallowly buried. Hauling the chest up, he brushed the soil away. Although the lid was carelessly nailed shut, he could pry it open easily; the nails were rusted and the lid of the wooden box had partially rotted through. Slowly, he realized what he was seeing. Bones. But not the bones of a horse, or a cow, or a dog. *Human* bones.[1]

"The news flew like wild fire," according to one newspaper account, "and people from many miles around visited the place." Rumor told that the grave in the field was at least ten or twelve years old and belonged to a "negro trader" from Georgia who had mysteriously disappeared after visiting the infamous tavern kept by Patty Cannon and her son-in-law Joe Johnson, located just over the Maryland line. For the previous several years, local authorities had become increasingly suspicious about numerous unexplained disappearances and other curious activities surrounding Cannon and her associates. Cannon and Johnson, meanwhile, had been embroiled in the highly profitable business of kidnapping free African Americans and slaves, holding them captive in the attic prison of their tavern or in the forests and swamps nearby, and selling them to slave dealers further south. In several recorded instances, unsuspecting slave traders were murdered for their money at Joe Johnson's border tavern, robbed, and their bodies buried secretly in the nearby fields. As it turned out, it was the discovery of the bones in the blue chest that led authorities to other burial sites—including the graves of several murdered children, one of whom was buried in Patty Cannon's garden—and, ultimately, to Cannon's arrest. Cannon never paid the full price for her deeds. She died just weeks later in the Georgetown jail while awaiting trial, an apparent suicide from a self-administered dose of poison.[2]

This latest news hardly surprised those who knew the border regions of North West Fork Hundred, Delaware: this landscape on the margins of Maryland and Delaware had developed something of a reputation. Described as "one of the most desolate and isolated points on an isolated peninsula," the area on the far northwestern side of southern Delaware's great Cypress Swamp was already well known for its relative lawlessness, fueled by the natural isolation of the surrounding communities, exacerbated by customary behavior and local reticence.[3] Patty Cannon, historically the most infamous resident of the area, had long used her tavern's location at the confluence of two states and three counties to evade authorities and had repeatedly enlisted geography as her ally by jumping from one jurisdiction to another whenever pursuit seemed imminent. Contemporary reporters of Cannon's capture and arrest marveled at the extent of her wickedness but also implied that local suspicions about her misdoings had simmered quietly in the area for years, and they suggested that the region itself bore at least part of the blame. "The neighborhood in which these terrible events occurred, the borders of Delaware and Maryland, has long been famous for negro stealing and negro trading, and 'Patty Cannon' and 'Joe Johnson' are familiar names to us," wrote one correspondent. "The people thereabouts were exceedingly ignorant and desperately wicked—but we hope that some improvement has latterly taken place."[4]

To contemporary observers, this area constituted a border region in more ways than one. This boundary area between Maryland and Delaware, located in the Nanticoke River drainage almost midway between the Chesapeake Bay and the Delaware River, differed substantially from other parts of the lower Delaware Valley in the late eighteenth and early nineteenth centuries. This part of Sussex County was more than just an isolated geopolitical and physio-

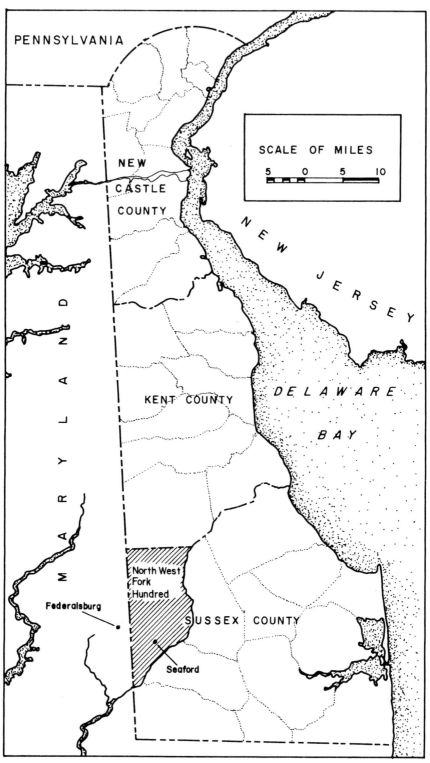

FIG. 3.1. North West Fork Hundred, Sussex County, Delaware

graphic edge zone that happened to shelter a band of outlaws. A landscape on the margins in material as well as conceptual terms, this was a frontier of sorts, a place where the earliest durable architecture bore closer links to the Chesapeake building traditions that prevailed to the south and west than those to the north and east, where a large influx of settlement occurred only *after* boundary disputes between the two states had been resolved in 1775, where much of the available land and wealth—even well into the second decade of the nineteenth century—was controlled by a tiny but powerful minority, and where patterns of slavery and land speculation shaped the boundaries of everyday experience. And, in the mental landscapes of contemporaries from outside the region, the area also operated as something of a cultural margin—a place inhabited by a rural forest society that was "not so much lawless as it seemed to be temporarily ungoverned,"[5] where the borders of civility and accepted comportment were often blurred, and where the landscape itself may have seemed to act upon the boundaries of everyday behavior.

Even as it was unfolding in the lower Sussex County countryside, the Patty Cannon story captured the public imagination. While contemporary accounts leave little doubt as to the veracity of the events themselves, the story surrounding those events quickly took on a life of its own. Little more than a decade after Cannon was arrested, the tale had attained all the dimensions of a local historical legend. In a somewhat embellished account that appeared only twelve years after Cannon's death, her story had already begun to blur the boundaries between historical fact and tall tale. According to this account, Cannon "died a most terrible and awful death." Once the poison she ingested took effect, "she raved like a maniac, tearing the clothes from off her body, and tore the hair from her head by the handfulls, attempting to lay hold and bite everything within her reach."[6] Through many such embellishments and repeated tellings, Patty Cannon's tale rapidly became an enduring part of Sussex County folklore, and "for generations after her death children are said to have quaked with fright and pleasurable terror at the sound of her name."[7] Even today, the story remains firmly entrenched in the region. Cannon's tavern is a notable spot on historic site lists, an official roadside marker attests to her notoriety, and local compilations of "haunted areas" and ghost walks seldom fail to mention her name. Patty Cannon is so infamous that three counties in Maryland and Delaware claim her as their own, and a modern subdivision in southern Delaware even bears her name.

Cannon's legend remains a compelling one, not just because of the violence and mystery surrounding the events that actually took place, but also because it suggests a great deal about the formation of a localized landscape identity. Many locales have their regional tall tales. These are often somewhat embellished stories about real people or retooled versions of major historical occurrences. Although such stories seldom represent events completely accurately, local historical legends typically contain a different kind of "truth," because they usually capture key features of the region or its history, often expressed as hyperbole. These legends also become locally entrenched for a reason. They represent place in a way that purely factual accounts cannot. They become important to a region not so much because they are entertaining stories but because they constitute a vital record of local history as it has been refracted through human responses to it.[8] In other words, they provide a unique kind of historical text by offering us a point of entry into the way people in a particular location may have thought about their world.

In his work on French cultural history in the early modern period, Robert Darnton has argued persuasively that the structure of French folktales provides a window into a past world view or *mentalité*. French peasants, maintained Darnton, found these stories "good to think with" by reworking them in their own manner, infusing them with meanings that resonated within their own personal cosmology. Such tales suggested how the world worked and provided their users with a critical strategy for coping with it.[9] William Cronon has argued more recently that, in all historical narrative, place matters. Moreover, as story and scene become intertwined, landscape and the way it is portrayed affect all historical narratives profoundly: landscape becomes as much a "character" in these narratives as any of the human actors. Cronon suggests that, because erasures in such narratives are just as significant as inclusions, where and when the narrator chooses to begin and end the story can also alter its meaning significantly. Thus, narratives about nature are innately "unnatural," and both human artifice and cultural significance stand "at the heart of even the most 'natural' of narratives."[10]

The Patty Cannon story has become important to the lower Delaware region largely because Cannon's persona, her actions, and the locational attributes that undergird her story carry tremendous symbolic freight within that border landscape. In short, like the economy, settlement, and political history of the area, the elements of the Cannon story constitute key components of the region's identity. Anyone who has attempted to sort fact from fiction in such a legend would agree that there are usually multiple versions of the tales. In the several versions of the Patty Cannon story, certain key elements persist. Violence, deception, and murder are the critical and often hyperbolic components of the plot line, to be sure, but the Cannon legend also revolves around a pervasive series of inversions and violations. Here were kidnappers who deliberately transposed freedom and slavery; here was a dwelling that functioned as both tavern and prison; here was a woman with the size and strength of a man; here was a mother who murdered young children. Moreover, this story always takes place on a very particular landscape stage. Here was a tavern that stood astride two states and three counties. Here was a gang that chained its captives on a small island in the Nanticoke and, under cover of darkness, spirited them to the Chesapeake Bay, where boats waited to take them to southern slave markets. Here, also, were outlaws who operated within the permeable margins between North and South, who called existing political boundaries into question by jumping from one jurisdiction and state to another, renegades who inhabited both regions, and neither region. And, in an era of growing concern with agricultural improvement, here was a clear violation of customary agricultural order: the appropriation of domestic storage furniture for coffins and the transformation of cultivated fields and gardens into secret burial sites represent clear cases of terrestrial *dis*improvement. Taken individually, each of the elements at the core of the Patty Cannon narrative—the violation of freedom, house, property, gender expectations, political jurisdictions, and the larger environment—represents a disquieting contradiction. Taken together, however, they suggest a mental landscape in which the blurring of traditional boundaries and the inversion of customary notions of property and order had, for whatever reasons, become resonant themes. In short, at least as portrayed in the Cannon tale, this region represented a borderland, occupying perceived as well as actual geographic and political margins. In this legend, then, this landscape on the margins had indeed become an actor, and place played just as important a role as Cannon herself.

The image of a conceptual and physiographic margin is hardly unique to this particular area and legend. Numerous other historical and fictional accounts of the area allude to a comparable intermediacy of the region and suggest the existence of the same kind of border landscape mentality that contemporary chroniclers implied. How, though, might such a pervasive regional leitmotif have originated? What, beyond the obviously sensationalistic aspects of the Cannon story itself, might have been the sources for such perceptions?

And, if this legend forms such a lasting and resonant part of this region's identity, how deeply were those landscape perceptions actually grounded in the material environment? While the Patty Cannon legend often straddles the boundaries between fact and hyperbole, it nonetheless offers us a useful point of departure for examining past physical as well as conceptual landscapes. Cannon's story, with its explicit inversions of property, freedom, and environment, suggests a great deal about the wider range of attitudes, behavior, and material conditions that defined the way people may have actually perceived—and occupied—this border landscape. And as intimated by Cannon's story and illuminated by a number of parallel contemporary accounts, this notion of a landscape on the margins offers a powerful metaphor for approaching the way this isolated rural society may have operated within the context of its material world.[11]

⁂

From our outsider's perspective nearly two hundred years later, "margin" provides a fitting way to describe this portion of lower Delaware. Physically and conceptually, "margin" connotes an edge region: the border or outside limit of a particular realm, perhaps, or an extreme limit beyond which something becomes impossible, or different, or is no longer desirable. "Margin," too, implies a place that both defines and is defined by a center; a margin cannot exist without a center to give it form. Yet "margin" can also suggest another kind of realm—a peripheral place where boundaries overlap, where ordinary distinctions become permeable, where differences become blurred and ambiguous, and where accepted notions of behavior and comportment begin to shift.

In geographic, physiographic, and material terms, the landscape occupied by North West Fork Hundred was a zone of transition in the early national period. North West Fork Hundred is located roughly at the midpoint of that ragged hangnail of Atlantic coast known now as the Delmarva Peninsula. This peninsula, which includes Maryland's and Virginia's Eastern Shores as well as the entire state of Delaware, extends about 170 miles from north to south and, with the Chesapeake Bay to the west and south and the Delaware Bay and River and the Atlantic Ocean to the east, is nearly surrounded by water. The peninsula's lower portion—that part encompassing Sussex County, Delaware, Virginia's Eastern Shore, and the middle and lower parts of Maryland's Eastern Shore—has long been recognized as a place that often differed from the rest of its three contributing states in terms of politics and culture, as well as geography. In the early years of the nineteenth century, the North West Fork part of Sussex County constituted a hinterland nearly as far removed from the urban commercial centers of Wilmington, Philadelphia, and Lancaster as it was from Baltimore. Like much of Sussex County, this area stood at the intersection of two major regional cultures: that of the Chesapeake to the west and south, and that of the lower Delaware Valley to the north and east. The region's cultural identity was affected, in part, by the origins of its settlers. While Quakers and Scots-Irish settled northern Delaware and engaged primarily in mixed-use grain and livestock farming, southern Delaware was settled by tobacco-growing, slaveholding migrants from the Chesapeake.[12] Environmental factors also contributed to the district's regional identity. Land throughout the county remained largely uncleared in the early national period, and local agriculture was based largely in corn, wheat, and swine, supplemented by extractive timber products such as building lumber, firewood, and shingles, often destined for the Wilmington, Philadelphia, and Baltimore markets.[13] North West Fork Hundred was economically oriented to these major urban markets, but in terms of the built environment, the area related more closely to Chesapeake traditions of impermanence.[14]

The area included part of the long-disputed borderlands between Maryland and Delaware, and it continued to be something of a political border zone through the late eighteenth century and into the early nineteenth century. Sussex Countians were so well known for their conservatism and strong Loyalist leanings during the Revolution that Congress expressed concern about "the spirit of Toryism" there and in neighboring parts of Maryland.[15] While settlers had claimed and farmed land in this area since the late seventeenth century, jurisdiction over the region remained contested until the state boundary was finally resolved in 1775. Extensive settlement was inhibited as a result. Many of the early plantations in the northwestern Sussex County area had been laid out on Maryland patents. By the time the boundary was settled in 1775, conflicting land claims in the region had become so complicated that much of the disputed area, including a good many of the old forest district patents in North West Fork Hundred, had to be resurveyed. The resolution of these longstanding disputes caused the Chesapeake influence in the area to become even more pronounced, since a significant portion of territory that had once been considered part of Maryland became part of Delaware. Even after these boundary disputes were resolved, borders appear to have been indefinite and land claims throughout the area remained uncertain through the end of the century. Although many of the more unsettled aspects of this rural society began to stabilize during the first quarter of the nineteenth century, the architectural forms favored by builders and their clients continued to reflect older traditions.[16]

In addition, while efforts toward land improvement and agricultural reform were taking hold in the more northerly reaches of Delaware, bringing more land into extensive cultivation, most of Sussex County remained forested, isolated, and unimproved even into the middle of the nineteenth century. Settlement was widely dispersed, towns were few, and much of the landscape remained in forest, scrub, and wetland. This countryside, which was dominated by cypress, Atlantic white cedar, loblolly pine, and oaks, and laced with swamps, hidden paths, and an occasional cleared farm, imparted a distinctive character to the region which contemporary descriptions often captured. Early historical accounts almost always labeled Sussex County farmers as "foresters," and more than a few late-eighteenth-century land surveys referred to the locations of the region's farmsteads as "in the forest."[17]

The region also fell at the edges of a physiographic divide: the densely thicketed central ridge which runs in a north-south direction through Sussex and contains almost all of the county's freshwater swamps defines the division between drainages. On the western side of this forested spine, the Nanticoke River and its tributaries lead southwest toward the Chesapeake, while in the eastern reaches, the Indian River drainage feeds the Delaware. These natural barriers have historically influenced the conceptual division of Sussex County between the eastern hundreds, which have looked to the Delaware River as their natural artery of commerce and link to the outside world, and the more western hundreds such as North West Fork, which looked toward the Chesapeake.[18] Still, even when the Chesapeake and Delaware Canal was completed far to the north in 1829, linking the markets of Philadelphia and Baltimore by way of Wilmington, geographic isolation and circuitous travel kept northwestern Sussex County far outside of the quickening mainstream. Just as in Northampton County on the Eastern Shore of Virginia, this relative isolation helped to determine early political and social relationships and played a significant role in establishing the region's sense of place.[19]

Finally, this area occupied another critical boundary zone: that between slavery and freedom for African Americans. With its substantial slave and free black populations, North West Fork and the surrounding lower Delmarva

Peninsula represented a kind of disquieting intermediate region that was any-thing but benign. Historian Barbara Fields maintains that, in terms of slavery and freedom, Maryland, especially its Eastern Shore region, occupied a po-litical as well as geographic middle ground. Like Delaware, Maryland exhib-ited a substantial free black and slave population in the nineteenth century, but the state's free blacks held a socioeconomic position that was deemed both essential and problematic. Maryland, argues Fields, "was a society divided against itself. There were, in effect, two Marylands by 1850: one founded upon slavery and the other upon free labor." The state's social and economic dif-ferences were also reflected in its politics. Northern Maryland was a largely white, free-labor society where industrialization had taken hold; southern Maryland was more than half black and represented "a backward agricultural region devoted primarily to tobacco." Maryland's Eastern Shore, however, the region that abuts the state of Delaware, represented a middle ground within the state. It was like the Western Shore's southern counties in that it was mostly agricultural rather than industrial; it was like the northern counties in its re-liance on cereal crops rather than tobacco. "Neither as slave and black as southern Maryland nor as free and white as northern Maryland," it occupied an intermediate position similar to that of Maryland itself within the Union. This unique position made the experience of slavery and freedom here nei-ther benign or mild. On the contrary, the close proximity of sizable free black and slave populations and the high volume of trade in slaves, Fields maintains, rendered the experience of enslavement here that much more bitter. Also, be-cause the slave regions were so close, kidnapping for sale out of state was prob-ably a constant cause of anxiety for free blacks. Thus, this "middle ground" position depended upon both the distinctiveness and the vigor of the North and the South. Fields contends that this position was not moderate, but in-stead, "parasitic," much like Patty Cannon and her cohorts, in that it fed off of the entrenched instability between two polities and social systems that would never be able to compromise.[20]

Similarly, Patience Essah has suggested that Delaware, and especially south-ern Delaware, also represented a political middle ground between slavery and freedom and that the distinct cultural difference between northern and southern Delaware likely planted the seeds for a prolonged political stalemate over state-legislated abolition of slavery. This antagonistic political, religious, and cultural difference, caused in part by differing economies and settlement patterns in the northern and southern parts of the state, gradually solidified into an important sectional boundary. Northern Delaware was largely Republican and Quaker and supported *de jure* emancipation; southern Delaware was mostly Methodist, Democratic, and rigidly opposed to abolition. Even contemporary observers found it persistently difficult to define the entire state as either "Northern" or "Southern." In 1787, James Madison linked Delaware to the North, but in 1859, Edmund Ruffin underscored Delaware's unusual position regarding the peculiar institution: "There are now thirty-three States in the Union, of whom fifteen only, if including Delaware, are slaveholding, or but fourteen, if excluding Delaware, which holds very few slaves, and is already, in sentiment and political action, almost identified with the more northern States."[21] Because Delaware lacked the room for expansion on its western frontier that most other states possessed, Essah argues, "politics in Delaware evolved into a 'house divided' between perpetually feuding cultures in its northern and southern regions."[22]

If the entire state of Delaware stood as a "house divided," the Sussex County area occupied a rather unique position within this larger border region. Sussex County in general and North West Fork Hundred in particular were strongholds of slavery within the state. In 1790 almost half of Delaware's slaves lived in Sussex County, and the proportion of slaves in several parts of the county remained high through much of the early nineteenth century. Moreover, as the Patty Cannon story attests, Sussex County was an especially fertile ground for kidnapping and selling blacks illegally, due to its convenient and permeable borders as well as the statutes prohibiting slave export. Fueled by the southern cotton boom and the abolition of the external slave trade in 1808, the demand and the price for slaves jumped markedly in the early nineteenth century, and kidnapping of free blacks and slaves for sale further south rose proportionately.[23] This isolated, forested, and geographically distinct region of Delaware occupied a racial and political border zone—a "middle ground" whose margins were continually being tested and reformulated.

References to the Sussex County region in contemporary accounts also suggest something of the conceptual space defined by this area. This was a region that formed a kind of cognitive border—a place that not only stood at several geopolitical and physiographic edges, but one that also existed on the borders of the imagination. By 1829, the year that Patty Cannon was finally arrested, the area's regional persona already appeared firmly entrenched, for the author of Cannon's story in *Niles Weekly Register* had not only made the somewhat environmentally deterministic claim that the inhabitants of this border region were "exceedingly ignorant and desperately wicked" but had also implied that neighbors tolerated, encouraged, or even partook of the types of behavior exemplified by the Cannon gang's misdeeds. This landscape on the fringes of Delaware and Maryland, this "neighborhood in which these terrible events occurred," had seemingly acquired its own unique identity, at least in the minds of some contemporary observers. The notion of the region as a sort of conceptual as well as physiographic margin constitutes a leitmotif that appears in descriptions of the physical landscape as early as the late eighteenth century and surfaces repeatedly throughout both historical and fictional accounts more than one hundred years later. In many of these period accounts, this area's distinct regional identity was one in which climate and landscape were often fused with culture and human actions. While some observers simply noted the distinctions between the Sussex area and other parts of the broader region, others implied that environmental factors, at least in part, worked to shape human behavior there.

In his 1792 *Account of the Situation, Climate, and Diseases of the State of Delaware,* William Currie attributed the unhealthy aspect of the Sussex County lowlands to the physiographic differences between this region and the rest of the state. Currie praised the atmosphere in the northern part of the state but described the low, flat, interior section as a disease-prone landscape plagued by "Marsh Miasmata and other noxious exhalations." Although the sea breezes and heavily forested portions of Sussex compensated somewhat for the unhealthful qualities of the low-lying ground, especially in more coastal regions such as the town of Lewes, to Currie, it was in "the inland part of the country, where heat and stagnation occur to exalt the noxious exhalations of our low grounds, that our state can truly be said to be sickly." Currie complained, "the east and South parts of this State are low and flat, and a considerable portion in an uncultivated condition, which occasions the Waters to stagnate; in consequence of which the Inhabitants are very subject to Intermittents and Remittents [fevers] in the latter end of Summer and beginning of Autumn." Seasonal fevers plagued the inhabitants of these inland areas, leaving them more susceptible to other diseases.[24]

When the duc de la Rochefoucauld-Liancourt visited Sussex County around 1800, he described a sparsely populated "country of slavery" that remained thickly wooded and largely uncultivated, due primarily to the greed and shortsightedness of wealthy forest proprietors: "four-fifths of the county of Sussex . . . remains yet uncleared. The woods are certainly in some places filled with water, but with little pains and expense nearly the whole of these grounds might be drained, and doubtless would be very productive; as all those that have been drained yield great crops."[25] Still, forest proprietors intent on maximizing their profits preferred to leave their lands unimproved. "The want of hands is an obstacle which prevents any attempts of this nature . . . and induces a number of proprietors of woods to believe that their ground is more profitable to them in its present state."[26] Woodland proprietors stripped their forests to supply Philadelphia with pine and cedar for a handy economic return. Even as late as 1851, after ditching and draining had improved a portion of the marshy Sussex countryside, contemporaries railed at the poor farming practices that still prevailed. Such practices, in the estimation of some observers, were largely based on greed. "I regret to say," wrote Sussex County resident Charles Wright, "that in a considerable portion of the county there has been, and still is, a great deal of bad farming; and so great is the desire of the owner of land to make it pay for itself by the lumber which can be cut on it to supply a city market, that he neglects the improvement of his land altogether. Land soon becomes waste and worthless by such means."[27]

An 1855 autobiographical account of a nearby section of the Delmarva peninsula by one of its most famous inhabitants likewise underscores this same notion of a "waste and worthless" landscape, imbuing the region with an almost palpable sense of dread. When famed abolitionist and former slave Frederick Douglass recalled his childhood home in Tuckahoe, Maryland, about 30 miles east of the Delaware state line, he described the same marginal quality that earlier observers continually noted. Tuckahoe, wrote Douglass, was a "singularly unpromising and truly famine-stricken district" that was "thinly populated, and remarkable for nothing that I know of more than for the worn-out, sandy, desert-like appearance of its soil, the general dilapidation of its farms and fences, the indigent and spiritless character of its inhabitants, and the prevalence of ague and fever." Douglass's suggestion that the barrenness of this landscape carried over even to its inhabitants exhibits the same kind of environmentally deterministic outlook that often suffuses observations recorded during Patty Cannon's time. To Douglass, at least, environment and regional character were intimately linked. The name "Tuckahoe," he wrote, came from the legendary "petty meanness" of one of its earlier inhabitants, who was guilty of "stealing a hoe—or taking a hoe—that did not belong to him. Eastern Shore men usually pronounce the word *took*, as *tuck; Tuck-a-Hoe,* therefore, is, in Maryland parlance, *Tuckahoe.*" The name stuck, and this place was "seldom mentioned but with contempt and derision, on account of the barrenness of its soil, and the ignorance, indolence, and poverty of its people. Decay and ruin are everywhere visible, and the thin population of the place would have quitted it long ago, but for the Choptank river, which runs through it, from which they take abundance of shad and herring, and plenty of ague and fever." Douglass described Tuckahoe as a "dull, flat, and unthrifty district" that was "surrounded by a white population of the lowest order, indolent and drunken to a proverb." To Douglass, this was also clearly a landscape of despair for slaves, "who seemed to ask, *'Oh! what's the use?'* every time they lifted a hoe."[28]

Douglass emphasized the importance of the region's isolation in attempting to explain why his own childhood experiences of slavery were harsher than the generally prevailing "mild" view of slavery as it existed in Maryland. While public exposure usually acted to temper the "cruelty and barbarity of masters, overseers, and slave-drivers," Douglass suggested that environmental seclusion bred evil. "There are certain secluded and out-of-the way places . . . where slavery, wrapt in its own congenial, midnight darkness, *can,* and *does,* develop all its malign and shocking characteristics, where it can be indecent without shame, cruel without shuddering, and murderous without apprehension or fear of exposure."[29] Although Douglass might not have agreed that physical environment determined culture or human actions, his implication here was clear.

Douglass was not the only chronicler to convey an image of an isolated, worn-out, barren, and dilapidated landscape that formed a hospitable stage for human corruption. This notion that geography somehow makes an impact upon culture and human behavior pervaded many contemporary accounts of the area, including fictional ones. George Alfred Townsend, who used the Sussex County landscape as the setting for his 1884 novel *The Entailed Hat,* set in the 1820s, during Patty Cannon's lifetime, effectively seized upon the "waste and worthless" quality of the lower Sussex County countryside as his narrative stage. The fact that Townsend's novel was written nearly sixty years after Cannon's arrest attests to this story's remarkable staying power and resonance within the region. As his protagonist Jimmy Phoebus nears the Johnson's Crossroads neighborhood where Cannon's tavern was located, the landscape is portrayed in unflattering terms:

Phoebus passed along the side of a large, black, cypress-shaded millpond, and found the boundary stone again, and took the angle from its northern face as a compass-point, and, proceeding in that direction, soon fell in with a sort of blind path hardly feasible for wheels, which ran almost on the line between the states of Maryland and Delaware, passing in sight of several of these old boundary stones. Not a dwelling was visible as he proceeded, not even a clearing, not a stream except one mere gutter in the sand, not a man, hardly an animal or a bird; the monotonous sand-pines, too low to moan, too thick to expand, too dry to give shade, yet grew and grew, like poor folks' sandy-headed children, and kept company only with some scrubby oaks that had strayed that way, till pine-cone and acorn seemed to have bred upon each other, and the wild hogs disdained the progeny.[30]

Using much of the same imagery that suffuses Douglass's autobiographical account, Townsend further described a social landscape that fell somewhere between lawlessness and passive complicity. This was a region where the long-running boundary disputes had thrown all land titles into question, where "instead of improving their lands, our voluptuous predecessors improved chiefly their opportunities," and where, because no man's property was certain, the unprincipled foresters of the Nanticoke "would rather trade with the enemy than fight for foolish ideas; and so this region was more than half Tory, and is still half passive, the other half predatory."[31]

Subsequent authors explored a similar motif, describing a landscape that existed somewhere on the margins of propriety and order. A 1926 regional novel by R. W. Messenger likewise seized upon the Cannon legend. Messenger, who also set his novel in the 1820s, likened the cultural persona of Seaford, Delaware, at the time one of the most populous towns in North West Fork Hundred, to that of Federalsburg, Maryland, about ten miles distant and only a few miles to the west over the state line. As Messenger's protagonist Hayward Tilghman ruminated about where he could possibly go for help in such a corrupt and isolated region, he offered a decidedly negative opinion of Federalsburg and its population based on his first contact:

Moreover, he had got a very bad impression indeed of the citizens of Federalsburg the afternoon before, when he had stopped at the hotel for a cooling drink and a bite of lunch. The landlord had been inquisitive beyond all sufferance. The whole appearance of the hostelry had been slatternly and uninviting. The loungers on the porch and in the bar room had talked of nothing but the size of the last herring and shad catch, which was just over; the inability of any of their citizenry to get the shirts off their backs without tearing them on account of the herring bones sticking through because of having eaten so many of that delectable fish; the recent activity of the kidnapping gangs in picking up stray blacks and how they wished them luck; the price of tan bark and the number of cords piled along the shores of the little tidal river and there waiting to be "scowed" five miles farther down the river to the schooners, which drew too much water to come any nearer than that to the town and which transported the tan bark to Baltimore for marketing.[32]

Messenger's imagery, with its emphasis on isolation, extractive industry, and the region's generally depraved population, recalls the comments of Townsend and Frederick Douglass from several decades earlier. Speculating on the reasons for the Patty Cannon gang's successes over so many years, the author went on to offer a sobering assessment of the entire region. To Messenger, Cannon and her gang had gained such a strong foothold in the area precisely because the surrounding area was so corrupt, and neighbors "who called themselves law-abiding" merely "winked at" the gang's blatantly illegal activities. Nearly one hundred years after the Patty Cannon story had become public, then, R. W. Messenger had alluded to the same conceptual landscape that the *Niles Weekly Register* correspondent had described back in 1829. Despite the lapse of time, there were certain aspects of this landscape that continued to resonate in the mental maps of outsiders. For, as Messenger wrote, "an extreme 'rottenness' existed to a greater or less degree over the whole section surrounding Johnson's Cross Roads otherwise such an abhorrent condition as actually obtained in the nest of villainy itself positively could not have been reached."[33]

In literature, this attitude toward the border region of Delaware and Maryland as both a conceptual and physiographic margin was doubtless fueled by the rumors and events leading up to Patty Cannon's arrest in 1829 and the subsequent entrenchment of that violent regional legend. But, as the fictional and historical accounts have intimated, the notion of a landscape on the margins arose from geographic and historical factors as well. While the "rottenness" to which Messenger, Townsend, and the *Niles Weekly Register* correspondent alluded was almost certainly more closely linked to the specific activities of Cannon's gang rather than to any characteristic embedded in the entire region, other physiographic and economic factors nevertheless may have contributed to a perceived regional *mentalité*. In the early national period, this Sussex County border region was a low-lying, heavily forested and sparsely populated "country of slavery" defined not only by the geographic isolation that these writers seized upon but also by uncertain property claims, profound socioeconomic inequality, and an extractive forest economy and environmental mentality that placed short-term economic gain above long-term land improvement.

Still, if this Sussex County landscape occupied a conceptual space "on the margins" in the minds of contemporary correspondents and subsequent writers looking from the outside in, might it have been experienced somewhat differently by its early nineteenth century inhabitants? After all, distinctions are often in the eye of the beholder, and a boundary may be perceived differently depending on one's vantage point.[34] Was this region a distinctive and definable "landscape on the margins" only to outsiders? Or was it "an insider's mental construct, a state of mind instead of a corner of a state, an area defined according to the natives' sense of place—their intimacy with their physical surroundings and with the distinctive political and economic circumstances and history of the region?"[35] How did the cultural, social, and topographical perceptions of outsiders relate to the way the inhabitants of this region actually experienced and occupied it? What kind of place is actually revealed in the material and documentary records, and how might the physical substance and textures of that landscape have informed the *mentalité* of its occupants—or, perhaps, the *perceived mentalité* implied by observers who interwove fact, place, and legend into the enduring stuff of regional folklore? For "sense of place," as Mary Hufford declares, "literally begins with the senses, with an ability to make sense of the environment, not only to tell what is there, but to understand the relationships between environmental elements."[36] In order to understand something of the reciprocal relationship between place as experienced and place as recorded, we must first turn to the landscape itself. Our comprehension of this particular early American place and others like it depends on a deeper understanding of the local society and their material world, and develops from a closer examination of the community and the way it occupied the landscape.

At first glance, the surviving landscape on this northwestern margin of Sussex County betrays little of its earlier past. The land itself—now mostly flat, open expanses of ditched and drained fields interspersed with patches of woodland—hardly fits de la Rochefoucauld-Liancourt's earlier description of a "thickly wooded and largely uncultivated" countryside. Only a few early buildings survive. Those few that do, like the Maston House (Figure 3.2), typically represent the life style of elites and, in their material of construction as well as their level of finish, are hardly characteristic of this region's landscape during the early national period. While the few remaining eighteenth and early nineteenth century buildings in North West Fork Hundred suggest the way a handful of elite individuals lived and generally demonstrate the area's close links to Chesapeake architectural traditions, they exemplify but a small segment of the buildings that once stood on the landscape.[37] To reconstruct this region's earlier physical and social environment, we must turn to documents such as tax lists, probate inventories, and orphans court records.

In economic terms, life in North West Fork Hundred was already startlingly inequitable by the end of the eighteenth century, and the boundaries between the very wealthy and the rest of the population were clearly delineated. This was a society in which the bulk of the taxable wealth was concentrated in the hands of a few. In 1801, one fifth of the population held nearly one half of the assessed wealth in the entire hundred. Just two years later, this same group

F I G . 3 . 2 . Maston House, Seaford vicinity, Sussex County, Delaware, photographed in 1995. Highly finished brick dwellings like this one were relatively rare in most of Sussex County and tended to be associated with the wealthiest residents.

controlled nearly two-thirds of the wealth, and by 1816, they held nearly *three-quarters* of the total taxable wealth, and the top 10 percent alone held well over half (see Table 7 in the Appendix).[38]

Land ownership structured this profoundly hierarchical society. In North West Fork Hundred, nearly two thirds of the population owned no land. Of those who did own land, just 5 percent controlled one-fourth of all the acreage. The tremendous gap that existed between rich and poor here was exacerbated by the extreme wealth of two brothers, Jacob and Isaac Cannon ("Cannon" was a very common surname in Sussex). The Cannon brothers ran a lucrative ferry operation on the Nanticoke River, controlled an extensive local money-lending business, and together owned nearly $40,000 worth of land as well as more than thirty stores, warehouses, tenant houses, rental properties, and many slaves.[39] In North West Fork Hundred, the wealth and power of the Cannon brothers dominated the local economy and pushed poorer folks even further toward the economic margins.

During their lifetime, the Cannon brothers exerted tremendous economic control and engendered a great deal of resentment due to their extreme wealth and uncompromising ways. According to one contemporary, "the Cannons were such a tyranical set of men, that no person could live by them in peace, all must submit to their dictation."[40] Between them, the Cannon brothers held over 5 percent of the taxable wealth in the hundred in 1816; that year, they were assessed for 32 tenant properties in North West Fork Hundred alone.[41]

When the brothers died only weeks apart—one of them murdered due to a business dispute—a contemporary diarist's caustic eulogy complained that the last thing that the "greate and rich Isaac Cannon" did before he died was to count his money. "I have known some of thier tenants to be oblidged to give them three bushels of corn in the fall, for one lent in the spring, to keep thier families from suffering; and then turn them out of house and home, after taking the last cents worth of property from them." He railed further:

> They stoped at nothing that they could lay hold by which they could make money. They gave a general credit. Of course their profits were enormous. They shaved paper and sold up every person that they could thus get hold of that was not able to pay them according to law. Hence they sold beds from under the sick, and in some instances the dying, took them away and left them to ly on the floor . . . They actually in some instances bought pots at constables sale in which

dinner of the poor createers were cooking and have taken off while boiling and emtied the victuals on the dung hill and left the women and children crying for want of food. Their whole lives has been one tissue of corruption, fraud and oppression. But they died in possession of a greate deal of wealth. Perhaps one hundred tenements, and hundreds of thousands of dollars, and except by their flatterers, as much despised by the community as any two men that have lived among us.[42]

Their "one hundred tenements" extended well beyond the boundaries of North West Fork, and included properties throughout Sussex County. Although the social and economic power of the Cannon brothers was unique to the North West Fork area, in the scale of their landholdings and the extent of their wealth they resembled a small but powerful handful of other wealthy landowners and forest proprietors in Sussex County, such as John Dagworthy and his son-in-law William Hill Wells of Dagsborough, who together exerted a similar degree of control over the Cypress Swamp forest districts.[43] Still, the Cannon brothers represent the extreme in this hierarchical forest society. Their remarkable wealth stands in stark relief compared to that of the bulk of the population and helps to illuminate the material worlds of others who also occupied this border landscape.

Property descriptions contained in Sussex County Orphans Court records suggest something of the organization of the region's economy and built environment. In the aggregate, these descriptions capture an architectural landscape that contrasted sharply with those farther north in the Delaware Valley, in terms of both land use and buildings. In particular, the tenuousness of the built environment that emerges from closer examination of these records throws the "marginal" landscape leitmotif that permeates the Patty Cannon legend and so many other narrative accounts into much sharper relief.

The orphans court assessments enumerated the property of individuals who died with minors and real property, and typically included detailed descriptions of all the buildings on the property, an assessment of their condition, a list of the tasks required to preserve the value of the property, and its expected annual valuation. Assessors also described and evaluated other improvements including gardens, fences, and orchards. The types of properties included in these assessments range from tenant farms and small parcels with few improvements, such as Morgan Williams's "three Hundred panel of Worm fencen all in bad repair and but one Apple tree and narey House of no Sort-

Whatsoever" to substantial tracts such as Curtis Jacobs's 500-acre tract outfitted with a substantial brick dwelling, peach, apple, cherry, and pear orchards, and nineteen separate outbuildings including a slave quarter, cookhouse, henhouse, loom house, stable, carriage house, stillhouse, and barn.[44] Since part of the assessors' role was to determine the projected value of a property for the benefit of a decedent's minor heirs, assessors were particularly attentive to physical appearance and condition. Still, orphans court records are not without their biases. Because these assessments dealt only with those individuals who owned real property, they typically recorded the holdings of only the wealthiest third of the population.[45] And since the gap between rich and poor was so extreme in North West Fork Hundred, this aspect of the valuations is revealing, especially as we examine the architectural landscape more closely. When linked with local tax valuations and probate inventories, these verbal descriptions can begin to intimate the contours of the region's built environment. An example of a single early-nineteenth-century farmstead and a general comparison with others in the region captures the broad outlines of this Sussex County architectural landscape and suggests something of the textures of this material environment.

When orphans court assessors appraised the estate of William Bradley on behalf of minors William, John, and Marina in 1808, they found a built assemblage similar to many in the area: "One Dwelling House Eighteen feet by Eighteen Two Story high not finished one old Milk House Nine feet Square in bad repair, one Smoke house twelve feet by Nine in good Repair, one old Cook House Eighteen feet by fifteen in Bad repair, one Corn House in bad repair, one Carriage house Twenty feet by Ten with one side Shedded in good repair."[46] Assessors also recorded a paled garden, a peach and apple orchard, and 1,780 panels of fence in bad repair. Bradley's house was cluttered with a predictable jumble of household goods including nearly two dozen pieces of furniture, kitchenware, stored food, and a loom and three spinning wheels.[47]

Although Bradley's house and holdings were, by local standards, quite substantial, the description of his farmstead exemplifies the way orphans court assessments represented most of the surrounding Sussex County landscape. Bradley's main dwelling house, all 648 square feet of it, probably contained a single, multipurpose, undifferentiated room on the first floor. These one-room-plan dwellings were typically accessed by an entry door opening directly into the main living space and were heated by an open fireplace. They usually contained at least one window, which was often set in the gable end away from the chimney or near the door, and a ladder or stair to a loft or upper story used for sleeping and storage. While larger one-room-plan houses like Bradley's rose a full two stories in height, much smaller examples, such as Philip Hughes's "one framed dwelling house 15 by 14 feet with a shed to one side," were not uncommon.[48]

William Bradley's farmstead typifies the appearance of much of the surrounding landscape. While Bradley's house seems small compared to modern dwellings, it represents a scale, type, and condition of housing that was quite common throughout the county as well as much of the broader region in the late eighteenth century and early nineteenth century. In Sussex County, nearly three-fourths of all families occupied houses of less than 450 square feet. Usually built of log or weatherboarded frame or, rarely, of brick, the average house typically contained four beds, two tables, one or two blanket chests, a corner cupboard, eight or nine chairs, one desk, and was domicile to a family of seven. One-room dwellings were so common in this region that, at the end of the eighteenth century, they housed roughly 85 percent of the inhabitants of southern Delaware and Maryland's lower Eastern Shore and were similarly widespread in southwestern New Jersey. These dwellings, which might house people from any socioeconomic group, were the dominant local dwelling type in Sussex County even into the last decades of the nineteenth century. Dwellings such as Henry Richards's "loged house Twenty feet by Sixteen with a wooden Chimney in midlin repair" and John Elliott's "framed house 22 feet by 16 in Very Bad repair" were typical in North West Fork Hundred. The tenant farmer who discovered the bones in the blue chest on that spring day in 1829 likely inhabited a one-room house.[49] Single-room dwellings could range from truly tumbledown hovels in sorry condition, such as Aaron Irons's "logg house 15 feet by 15 much Racked with the late Gale of Wind," or Manlove Adams's "old logged house fourteen by Sixteen with an earthen floor and in bad repair," to houses with finely finished interiors containing expensive furniture.[50] William Bradley, for example, stood in the wealthiest tenth of the population in North West Fork Hundred, most of whose houses nevertheless contained less than 450 square feet on the first floor, were built of log or frame, and exhibited varying levels of architectural finish ranging from clay-daubed wooden or puncheon chimneys, earthen floors, and exposed log walls to weatherboarded frame exteriors and finely executed ornamental brickwork.[51]

Larger and more elaborately finished and furnished dwellings also stood throughout the region, but they were far less typical. Frame dwellings with brick gables, commonly found farther west in Maryland but extremely rare in Sussex County, appeared occasionally in North West Fork Hundred and demonstrate that area's close ties to Chesapeake architectural traditions.[52] When, as the result of a violent storm in 1776, David Polk's 20-by-40-foot frame house was "tore to pieces, and has become a Wreck," the orphans court decreed that the guardian was to build from the salvageable pieces "a new house twenty feet long and eighteen feet wide with one brick Gable end, with one fireplace below two doors & two windows below under pined with brick the doors and window shetters to be well hung with Iron Hinges, and a Good Coat of tarr, with a good plank floor above and below." Trustin Laws Polk's frame plantation house was similarly constructed, "with a Brick Gavel [gable] end," and Constantine Cannon's and Levin Tull's frame houses were also built with brick gables in the Chesapeake fashion.[53]

Some houses were exceptional not so much for their method of construction but instead for their size and furnishings. Saxagotha Laws's "Large Dwelling House" eclipsed the one-room-plan houses in the area. It had five rooms on each of two floors and stood "in reasonable Good repair." Henry Hooper's brick house included the usual complement of household furnishings typical of the area but also contained luxury goods such as looking glasses, a walnut candle stand, and a tea chest. Philip Hughes, the wealthiest person in North West Fork Hundred in 1803, owned household goods that clearly linked him with Sussex County's rural elite; they included a mahogany dining table, a card table, an inlaid tea table with a full complement of tea equipage, two looking glasses, several books, a map of Delaware, and progressive agricultural machinery including two Dutch wheat fans. Still, dwellings and estates like these were unusual, and houses far less substantial than Bradley's were the rule throughout this border region of Sussex County.[54]

William Bradley's farm was much less substantial than many farm complexes in northen Delaware and southeastern Pennsylvania, but it eclipsed most others in its immediate neighborhood. With its two-story dwelling, milkhouse, smokehouse, cookhouse, carriage house, and paled garden and orchards, Bradley's plantation constituted a fully developed agricultural complex in the larger Sussex County landscape. The large number of outbuildings was unusual in an area where few farms contained additional buildings of any sort and where pigs, sheep, cows, and oxen were often left in outdoor pens or permitted to roam freely through the forest. And, although Bradley's corn, milk, and cookhouses remained "in bad repair," his dwelling, smokehouse, and carriage house were in good condition—an unusual circumstance in a countryside where the majority of buildings recorded by orphans court assessors were described in such terms as "midling tolerable," "bad," "wanting," "sorry," "wretched," or "falling down."[55]

Conditions of most buildings in Sussex County generally strained the limits of the assessors' accepted maintenance standards. The tenuous and somewhat disorderly architectural environment that emerges from the documentary record also recalls the "marginal" landscape leitmotif that was such an enduring aspect of factual and fictional accounts of the area. Because orphans court assessors were charged with listing the current and projected valuations of the properties they recorded, they paid special attention to the physical condition of the buildings and farms they saw, often noting decay or damage and recommending repairs. Yet maintenance of properties was often neglected. Farmsteads recorded throughout Sussex County stood in various stages of deterioration and rot, and many were inhabited in almost ruinous condition. William Polk's chair house and smokehouse both stood "in tolerable repair except the glass in the North West end of the large House" was "much broken in a Hail Storm." Joshua Hays's old dwelling house and smoke house remained in poor condition in 1829, and assessors claimed that they both "wants Weather boarding and We consider Will take from twenty to twenty five dollars to repare the same." In 1795, orphans court appraisers recommended several repairs to Peter Waples's house, complaining that "the front and the back dores of sd. dwelling House Are so brokin as to need repairing" and that "There are two Chimneys belonging to said Dwelling House One of wich is necessary to be takin down and Rebuilt as it a good deal sloops and some what likely to faul." Recording the inexorable process of decay, assessors remarked that Obediah Smith's 22-by-18-foot frame house was "tolerable well finished of, but now growing old," that the roof of one of John Shepherd's "two old logg's Houses" was "tumbling in," and that Allin Short's log kitchen stood with "the ruff gon" and "the logs old." And Uriah Brookfield's 25-by-25-foot house "with a Piaza out of repair" resembled the congeries of dilapidated outbuildings and rotten fences cluttering the rest of his Sussex County property, including an old milkhouse and smokehouse both "very much out of repair," a crib "without a ruff," 160 apple trees "rather old and some nearly dead," and an old lumber house "with a brick chimnie just tumbling down."[56]

No matter whether they were ramshackle, "midling tolerable," or, rarely, brand new, though, outbuildings of any sort still tended to be the exception rather than the rule. As in much of Sussex County, farmsteads in North West Fork Hundred included few additional domestic support structures or farm outbuildings, in contrast with more northerly farmscapes in Lancaster County, Pennsylvania. Most properties consisted only of a dwelling and perhaps one additional building, which was most likely to be a meathouse or smokehouse, cornhouse, or kitchen. Farmsteads like William Bradley's, with full ensembles of support buildings, usually belonged only to the wealthiest individuals.[57]

Domestic outbuildings, when they were present, usually fell into one of the three broad functional categories of food preparation, artisan work, or textile production. Only about a third of all properties were equipped with separate outkitchens and cookhouses. Appraisers surveying Hughitt Layton's farmstead found a domestic landscape that typified much of the region: some indifferent fruit trees and fencing and an old wattled garden, a frame house "in tolerable repair," and "One Old Kitchen very Sorry." More finely finished kitchens such as Henry Hooper's "Hue'd log Kitchen with Brick Chimney 16 feet by 16 in Good repair" were uncommon. As with other buildings in the region, many kitchens bordered on ruin. David McElvane's "Old Fram'd Kitchen" was "much Out of Repair the Ruff thereof being much worn is become leakey and the Chimney obliged to be Supported by a Prop."[58]

Outbuildings such as meathouses and smokehouses, milkhouses, and corn cribs appeared about as infrequently as outkitchens. About one-half of all the properties in old North West Fork included a meathouse or smokehouse used for smoking, curing, and storing pork and beef. Many of these windowless, gable-roofed buildings resembled Christopher Nutter's 10-by-10-foot "Smoak house almost Rotten." While brick smokehouses were occasionally built, such expensively constructed outbuildings were rare. Most local farmers constructed their smokehouses with log or frame walling.[59] Similarly, milkhouses or dairies like the one on William Bradley's farm occasionally shared the farmscape with the main dwelling. They tended to be small frame buildings, and the more ephemeral ones, such as Nathaniel Horsey's "one milk House standing on four posts four by three & ½ feet in midling repair" and Whittington Williams's "one framed Milk House 8 by 8 & 1 Do. on Posts 4 by 4," were clearly constructed so that they could be moved easily, recalling Chesapeake traditions of impermanence.[60] Cornhouses like Bradley's, also known as corn cribs or stacks, stood on around one third of all properties. Like most other Sussex

County outbuildings, these small gable-roofed log or frame buildings often appeared ramshackle or nearly ruinous to assessors. Nathaniel Horsey's "one corn stack in good repair" and Obediah Smith's "One Good Hude log'd Corn House large" were far less typical than Esther Lednum's "one old corn house of logs ten by twelve in very bad repair" or Uriah Brookfield's "one crib without a ruff."[61]

Barns, which formed the centerpieces of so many substantial farms farther north, were notably absent in this border landscape, suggesting a rather different agricultural emphasis. Despite his relative wealth, Bradley did not own a barn. Similarly, in sharp contrast to the more intensively cultivated landscapes of Lancaster County, most Sussex County farmscapes usually lacked any barn whatsoever. Barns stood on less than one-fifth of all recorded properties in North West Fork Hundred and were equally uncommon throughout the rest of the county. Of the barns that assessors did describe, most averaged around 20 by 30 feet, were one story high and covered with a gable roof, were built of frame or log, and, like other Sussex County barns, were smaller and less developed than barns throughout the rest of the lower Delaware Valley region.[62] And, following the pattern of the dwellings and other outbuildings throughout this landscape, most of the barns stood in "midling," tolerable," "sorry," or "bad" condition. Orphans court assessors decided that Jonathan Jacobs's frame barn, "in very bad repair," needed extensive repairs to set it right. Similarly, the appraisers recommended that Alexander Laws's log barn be weatherboarded and chinked. Like many other buildings on the Sussex County landscape, barns were also likely to be moved from one place to another or even reappropriated from structures originally designed for another purpose.[63]

Even though Bradley owned a horse and colt, his farm also lacked a stable. Only around one-quarter of all properties enumerated in North West Fork Hundred included a stable, and most of these—small gable-roofed log or frame buildings, like Hudson Cannon's "4 slab stables Sorry"—languished in poor condition. Fewer than half of the taxable population owned horses in 1816; of those who did keep horses, most owned only one or two.[64]

While William Bradley's farmstead was thus exceptional in the number of additional buildings it contained, the types of outbuildings represented were typical. Farmsteads in North West Fork were not without other outbuilding types; they included artisan shops, slave quarters, and even more functionally specific structures such as poultry houses, cider houses, loom houses, or still-

houses, but these buildings were comparatively rare. Commercial and industrial buildings such as storehouses, grist mills, and complexes such as David
Pennewell's fully equipped tannery replete with currying shop, shoe shop,
bark and beam house, and thirty-four tanning vats also appeared occasionally,
but these, too, were uncommon.[65]

Also, while Bradley was clearly one of the more prosperous farmers in the
area, unlike many of the wealthiest inhabitants, such as the Cannon brothers,
he did not own additional, tenant properties. Land, as we have seen, belonged
to only the wealthiest few. In this respect, too, the Sussex County landscape
differed from the fertile rolling countryside of the Lancaster Plain farther
north. In a countryside where nearly two-thirds of the taxable population
owned no land to speak of, tenancy constituted an ongoing reality of this hierarchical border landscape. Like the tenant farmer whose discovery of the
bones in the blue chest led to Patty Cannon's arrest, most people worked
rented land and occupied houses on someone else's property. Some of these
additional dwellings were located adjacent to the main "mansion" house,
while others, such as William Polk's "small dwelling House in the Corner of
One of the fields on the South East side of the Tract" or Obediah Smith's
"log'd Housen on one end of said Plantation" stood some distance away. Like
most primary dwellings, these secondary houses were usually small, built of
frame or log, and poorly maintained, often echoing the appearance of the
main dwelling. Outplantations—physically separate land parcels owned by a
single individual—were also often worked by tenants and enumerated as separate tracts; they also usually contained additional dwellings as well as an outbuilding or two. Like the "greate and rich Isaac Cannon," a few of the wealthiest individuals maintained several separate tracts of land, many of which were
occupied by tenants. For example, when orphans court appointees assessed
Trustin Laws Polk's property in 1796, they listed nine separate pieces of property in addition to the dwelling plantation, seven of which contained dwellings
occupied by tenants.[66]

Fencing, such as the garden pales and thousand-odd panels of field fence
that surrounded William Bradley's land, also constituted an important aspect
of this environment. From Nathaniel Horsey's simple wattled garden fence to
Charles Dean's "two thousand One hundred and fifty three Pannels of fence
in tolerable repair," fencing divided, defined, embellished, and protected
property and appeared in almost every orphans court description in North
West Fork. As in the architectural environment, however, maintenance of

these boundary markers was often deferred. No matter what their condition, household yards, outfields, and orchards were defined or surrounded with open post and rail, worm, or post and plank fencing. And, to protect their contents from hungry predators and roaming livestock, gardens were typically enclosed with fences constructed of closely spaced pales, woven wattles, or withes interwoven with slabwood, rails, boards, or upright pickets.[67] Roaming livestock constituted a perpetual problem. While stables and barns sheltered some farm animals, others wandered freely, wreaking havoc on open or poorly fenced gardens and fields. Property assessors, in fact, occasionally took note of roaming as well as penned livestock. Inventory takers enumerated Robert Williams's sow and six pigs "as they run." And inventory takers recorded David Pennewell's goods and livestock as including "14 shoats as they run in the road (wild)."[68]

Because livestock required a considerable financial investment, only the wealthiest individuals, like Bradley, could keep large numbers of animals. William Bradley's sizable collection of horses, cows, pigs, and sheep was well above average. In 1816, nearly half of the population of North West Fork Hundred owned no livestock at all, and most farmers owned only a few animals. Cows and horses were deemed the most essential farm creatures. Individuals who owned but one or two animals were most likely to own a single cow or a horse, or perhaps one of each. Those who could afford a little more usually chose to keep a few more cows, a few pigs, and perhaps another horse or a team of oxen for the heavy work of the farm. Typically, farmers in North West Fork kept sheep only if they already owned the basic complement of a half-dozen or so of the more essential farm animals, including a horse or two, a cow or two, a few pigs, and perhaps some oxen.[69]

For most farmers who had more than a couple of animals, like William Bradley, livestock constituted a significant portion of their estate. For example, over one-quarter of Manlove Adams's probated estate consisted of livestock, including three horses, two colts, eleven cows, a yoke of oxen, and fifteen sheep, and the animals listed in David Richards's probate inventory reflected a similar investment strategy.[70] In fact, investment in farm animals and farm equipment often eclipsed by far the amount of money tied up in household goods. William Bradley's modest investment in household furnishings was typical: while livestock and farm equipment represented around one-fourth of his monetary worth, his furniture, kitchenware, and household goods constituted only around one-tenth.[71]

Most of Bradley's wealth—like the wealth of so many other others in North West Fork—consisted of property that would contribute to the overall success of his farm: livestock, farm equipment, and, most importantly, slaves. Bradley's five slaves constituted a significant portion of his probated estate. As we have seen, in the early nineteenth century, slave ownership in North West Fork Hundred, where between one-third and one-quarter of all households kept between three and four slaves, was among the highest in Sussex County. This high proportion of slave ownership recalls La Rochefoucauld's observation of this border landscape as an isolated, sparsely populated "country of slavery" and underscores this area's close connections to nearby Chesapeake landscapes in which slavery played such a significant role. The pronounced presence of slavery not only defined an important aspect of this forest community, but it may also have fueled the rapid entrenchment of the Patty Cannon legend and contributed to perceptions of the region itself as a cognitive border zone.

In 1800, one-third of all North West Fork Hundred households owned slaves; in 1810, slave ownership was higher than anywhere else in the county, and in Sussex County as a whole there were almost twice as many slaves as in either of Delaware's other two counties. Although slave ownership declined countywide in the first two decades of the nineteenth century, the percentage of slaveowning families in North West Fork remained high. (See Tables 8–13 in Appendix.) Historians have suggested several reasons for the persistence of slavery in this area. They include the region's distance from cities, its isolation, its relative economic backwardness, and the absence of a strong Quaker presence or the kind of antislavery activism that often prevailed farther north. Furthermore, as the Patty Cannon story attests, the proximity of neighboring slave states made it a relatively simple matter for slaveholders to dispose of surplus slaves by selling them south.[72] The persistence of slaveholding into the early nineteenth century may also reflect the ongoing maintenance of established customary practices. Slaveholders listed in later tax assessments were typically the children of the preceding generation of slaveowners, and wills and probate inventories commonly enumerate individual slaves as legacies for the decedent's heirs. For example, while Hughett Cannon willed his land and house to his sons Wingate, Daniel, and Elijah, he bequeathed to daughter Polly "one negro boy named jacob also one negro girl named Rachel," both of whom were infants at the time, and to his wife Sally he willed Dinah, who was then 44 years old.[73] In short, although slaveownership in general was declining in the early nineteenth century as manumissions increased, the older practice of slaveholding still remained firmly entrenched in this region, especially within existing white kinship networks.[74]

While just under one-fourth of all North West Fork households still included some slaves even as late as 1820, buildings enumerated as "quarters," "cabins," or "negro quarters" seldom appear in orphans court records throughout the period. Some slaves doubtless occupied purpose-built quarters. Trustin Laws Polk's sixteen slaves may well have resided in his "frame quarter with brick chimney in bad repair," and some or all of Levin Cannon's eleven slaves, including 43-year-old Peter and Rose, Rose's 3-month-old child, 22-year-old Plymouth, teenagers Stephen, Savile, Mina, Mint, and Tamor, 5-year-old Peter, and 3-year-old Beavous, may well have lived in his old 12-by-10 log quarter. Still, the slave population in North West Fork far exceeded what could have been accommodated in the number of quarters listed in the documentary records. The vast majority of slaves most likely lived elsewhere—in plantation tenant houses, outbuildings, lofts, or other secondary spaces.[75]

As important as slavery was to the area, the presence of large numbers of free blacks was also a critical aspect of this region's identity, for within this landscape of slavery there flourished a significant population of free African Americans, most of whom labored as artisans and small farmers.[76] By 1820, the total African American population in North West Fork was about evenly divided between free and slave, but by 1830, free African Americans constituted nearly one-fifth of the total population. Nonetheless, relative to other parts of Sussex County, the enslaved population of North West Fork remained high. The presence of this large free black population thus echoes the kind of demographic situation in Maryland discussed by Barbara Fields, for it was the very presence of this large free black population within such a "landscape of slavery" that carried such resonance in depictions of that environment.

The significant population of both slaves and free blacks remains one of the most provocative juxtapositions within this border landscape and again brings to mind the inversion of freedom and bondage that the Patty Cannon story so disturbingly illustrates. In fact, several contemporary accounts seemed to pay special attention to this region's potential for blurring the boundaries between bondage and freedom, often linking aspects of the landscape with human behavior. In a letter dated January 20, 1827, Joseph Watson, then the mayor of Philadelphia, commented on the activities of the Cannon-Johnson gang "whose haunts and head quarters are now known to have been, on the dividing line between the states of Delaware and Maryland, low down on the peninsula, between the Delaware and Chesapeake bays." Watson went on to underscore how being on the border between political jurisdictions *and* the presence of significant free black and slave populations made this location convenient for the gang's activities. "The local situation of the country afforded them great facilities in carrying on this most iniquitous traffic, the bond and the free, have been equally subjects of their rapacity; numbers of slaves have been stolen from Maryland and Virginia, and carried to the southern and more western states for sale."[77] Similarly, an 1822 account alluded to the same racial twilight zone that other contemporaries described:

> In Maryland and Delaware, the same drama [that of the seizure and sale of African Americans into slavery] is now performed in miniature. The arrival of the Man-Traffickers, *laden* with cash, at their respective *stations*, near the coasts of a great American water, called justly, by Mr. Randolph, "a Mediterranean sea," or at their several *inland posts*, near the dividing line of Maryland and Delaware, (at some of which they have grated prisons for the purpose) is the well known signal for the professed *kidnappers*, like beasts of prey, to commence their nightly invasions upon the *fleecy flocks;* extending their ravages, (generally attended with bloodshed, and sometimes murder,) and spreading terror and consternation amongst both freemen and slaves throughout the *sandy regions,* from the western to the eastern shores.[78]

In short, just as la Rochefoucauld had intimated, this was surely a "country of slavery" as well as one of freedom, but, for most African-Americans inhabiting this landscape on the margins in the early nineteenth century, it was also unquestionably a border zone. Entrenched local patterns of slaveholding, the growing value of chattel labor to the south and southwest, the large population of free blacks, and a generally isolated location within a sparsely settled and largely forested area combined to create hospitable conditions for the kidnapping activities of Patty Cannon's gang. In the mental landscapes of contemporaries and subsequent chroniclers, an environment that constituted such a geopolitical and physiographic margin and simultaneously supported such large populations of both slaves and free blacks provided the likes of the Cannon gang ample opportunity to violate established boundaries of property and propriety. While stories of Underground Railroad outposts operating throughout the region abound, the much-touted presence of the Cannon gang's "reverse Underground Railroad" in the same location must have constituted a disturbing and nagging reminder that margins could, unfortunately, be permeable in both directions, and that the potential to be transformed in both positive and negative ways remained ever-present.

✛ ✛ ✛

As we have seen, William Bradley's farm represents one of the finer plantation complexes in Sussex County. Yet his situation hardly seems ideal compared to other landscapes in the lower Delaware Valley. The improved landscapes of the Delaware Piedmont, southeastern Pennsylvania, and parts of southern New Jersey contrasted sharply with the farmsteads and outfields recorded by orphans court assessors in the forested Sussex County countryside. In southeastern Pennsylvania and northern Delaware, for example, the large, multipurpose bank barns that consolidated crop storage, stabling, and work functions under a single roof formed the working centerpieces of many farmsteads. On these very different landscapes, large agricultural buildings like the "overshot" barns that Warwick Township's tax assessors described in 1815 multiplied significantly in the early nineteenth century. Farms in many of those areas also typically held proportionately more acreage in production, and houses were likely to be larger, more substantially built, and divided into multiple rooms. The origins of many of the Sussex County region's earlier inhabitants, like those of the Pennsylvania German areas, affected the way the landscape took shape. Although some early settlers in the county emigrated from elsewhere in the Mid-Atlantic, much of southern Sussex County was settled by people who migrated northward into Maryland from Northampton County on the Eastern Shore of Virginia.[79]

Sussex County's rural economy also differed notably from other parts of the lower Delaware Valley and remained distinct from areas where wheat cultivation, dairying, animal husbandry, and the early stages of industrialization had begun to take hold.[80] Farm products listed in Sussex County estate inventories usually spanned a fairly predictable range, focusing primarily on a mixture of corn, flax, wheat, occasionally rye, and mixed forest products such as tanbark and shingles.[81] While farming clearly factored into the local economy, timbering for the Philadelphia market, as La Rochefoucauld-Liancourt and other voices remind us, dominated in 1800 and remained a driving force in the region even at midcentury. This lumber-based extractive economy not only transformed much of the arable land into "waste and worthless" terrain but also spawned a remarkably inequitable society in which, at least in the first part of the century, the gap between individuals like the "greate and rich Isaac Cannon" and everyone else, and between African Americans and the rest of the population, continued to widen.

Land use in this forested countryside also contrasted sharply with the more improved farmscapes to the north. Relative to the rest of the lower Delaware Valley, the amount of land under agriculture was comparatively small. It was not uncommon for farmers to work only 10 percent to 25 percent of their land, leaving the rest uncultivated or in timber. In 1816, almost half of the land in North West Fork Hundred remained wooded, whereas in parts of Lancaster County and much of southwestern New Jersey almost 100 percent of the land was improved by 1815. Just as significantly, tenants worked much of the Sussex County acreage.[82] These farmscapes even stood apart from those of northern Delaware, where between 80 percent and 90 percent of the land was in tillage, and wheat, cereal crops, and dairying predominated.

The improved farmsteads of the rest of the lower Delaware Valley throw this forested, "waste and worthless" Sussex County landscape into sharp relief. Similarly, the way this particular border region was captured in the Patty Cannon legend and fictional accounts resonates in numerous ways with the historical and material record, suggesting a range of attitudes, behavior, and material conditions that remain as distinctive as Patty Cannon's story remains unsettling. At the heart of these resonances between the hyperbole of the legend and the physicality of the material world lay a series of very real juxtapositions which characterized this border region: freedom versus bondage, extreme wealth versus significant want, open or "improved" tracts versus vast expanses of closed forest. This was an isolated and ambiguous landscape, one that simultaneously stood largely outside of the economic mainstream but that was still enmeshed in the transactions of a larger economic world, one that was caught between the Chesapeake regional culture to the west and south and that of the lower Delaware Valley to the north and east, one that was both like and unlike the landscapes to the north and south.

In the mental landscapes of contemporary observers as well as in many socioeconomic, cultural, and material ways, this Sussex County region occupied a place on the margins and one where landscape and human interactions intertwined in complex reciprocal relationships. In our efforts to make sense of the past, we need to remember how compelling this sense of place can be. History, after all, has a corporeal aspect—every event occupies a physical dimension, and all actions are ultimately grounded, one way or another, in the landscape. Places, which possess their own geography, natural history, and embedded perceptions, not only ground the physicality of historical events—they also can constitute both actor and stage.

# 4 CHAPTER

# *Mapping the Ancestral Landscape*

Beneath a leaden winter sky in 1888, a photographer frames another piece of the flat southern New Jersey landscape through his camera lens. An old two-story gable-roofed brick house with an attached brick addition fills the center of the lens (Figure 4.1). Ragged leafless trees stand scattered across the facade. A tidy horizontal board fence in the foreground defines the front yard; a tattered, less formal fence tumbles off to the right. Just behind the front fence stands the blur of a small human figure in motion—a child, perhaps— and beyond the house, the wooden gable of a barn slips past the edge of the frame to the left. But upon closer examination, this dwelling presents an odd public face in the photograph. While its gable is nearly blank save for a decorative chevron of contrasting blue-glazed bricks near the roofline, its facade presents a more uneven appearance. Irregular door and window openings lend a curiously gaping look. And the opposite elevation, not even visible in this photograph, is actually the house's most formal side, with much more balanced fenestration and ornamental red- and blue-glazed Flemish bond brickwork. Yet, because eighteenth-century water routes have since silted in and nearby roads have been realigned, this elevation now faces, not a public road or a landing on the nearby creek, but an open field bordered only by an uneven tree line. Changes in transportation routes transformed the original back of this building into its most public side, relegating its original front, now visible only from the rear yard and the farm field beyond, to a much more private landscape.

No matter, for this more irregular elevation is the front—the public side— of the house now, and the one that is most appropriate to capture in a record as lasting as a photograph. Working deliberately, the photographer freezes the setting on his glass plate and adds it to his growing stack of images. His other photographs, while taken of different buildings and in different places in the surrounding region, tell a similar story: all are houses—old houses, typically at least one hundred and fifty years old or more by then—most are brick and two stories tall, most are associated with the founding Quaker families in the area, a good many were built by local Quaker brick artisans, and many, like this one, have blue brick patterns in the form of chevrons, zigzags, dates, or initials braided into their otherwise blank red gables.[1] Most tellingly, all are already minor landmarks of sorts: common physical points of reference sedimented from an earlier landscape; "old" places that help the inhabitants re-

FIG. 4.1. William Hall house, Mannington, photographed in 1888. The orientation of the house was changed when primary transportation routes shifted from water to land. The most formal and ornamental side of the builing, not shown here, faced the water. Photograph by Thomas Yorke; courtesy Salem County Historical Society.

call the region's past; surviving artifacts that effectively map, both physically and conceptually, a part of the settlement history of the southern New Jersey countryside.

The house was the William Hall Jr. house, built in Mannington Township in 1724 by a descendant of a prominent local family; and the photographer was Thomas Yorke, who in the late 1880s set out to capture on glass-plate negatives the earliest eighteenth-century buildings remaining in the South Jersey landscape. While many of the buildings Yorke photographed have since disappeared, several still stand today, scattered throughout Mannington and various parts of Salem and Cumberland Counties on New Jersey's inner coastal plain. Yorke's collection of images survives today in the form of a thick photograph album resting in the Salem County Historical Society. The images show houses in their surrounding yards and landscapes, sometimes people, occasionally animals. Each photograph is identified in Yorke's neat handwriting by location and sometimes by owner; in addition, most of his views contain genealogical or anecdotal information pertaining to the history of the houses and the early families associated with them. Yet Yorke's photographs have served more than a straightforward documentary function. By linking these houses to their family genealogies, Yorke simultaneously glorified the physical remnants of the colonial landscape even as he bound these remnants to the elite Quaker families who built them.

Still, Yorke was not the first to monumentalize the region's early built landscape. By immortalizing southwestern New Jersey's earliest surviving brick houses through the then-young medium of photography and by linking those photographs to family genealogies, Yorke was following the lead established in the previous century by the region's elite founding families, those who built these houses. Prominently initialed, dated, and ornamented on their most public elevations, the patterned brick dwellings of southwestern New Jersey were ancestrally linked from the moment they were built. From the earliest decades of the eighteenth century, these family monuments proclaimed the lineage and property of the area's most influential inhabitants even as they provided durable navigational landmarks in the "sea" of this flat tidewater landscape. Yorke merely reaffirmed and historicized a landscape conception that had been established, codified, and firmly monumentalized almost two hundred years earlier.

At around the same time, and likely in response to the overlap of the nation's centennial with Salem's bicentennial, others were following the route mapped by Yorke. Just a decade before Yorke carted his camera around South Jersey, Thomas Shourds had written his *History and Genealogy of Fenwick's Colony,* in which he compiled several individual family genealogies and noted where these families first settled. But Shourds dealt with more than just genealogy and settlement history. Houses—particularly "ancient" houses—salted his genealogical accounts. Throughout his book, Shourds repeatedly betrayed his reverence for the region's venerable ancestors as well as the "ancient" buildings they erected, and he continued to link these ancestors to surviving eighteenth-century dwellings, typically listing the names of the current owners in a conscious effort to link present occupants with past landscapes.[2] While Yorke was photographing "ancient" houses, J. H. Simkins, another photographer by trade, was producing a series of miniature paintings of some of the same buildings. Although Simkins painted several early buildings, his images focus primarily on the patterned brick houses. And although he was a practicing photographer, he chose for unknown reasons to immortalize these architectural monuments through the medium of painting rather than photography.[3]

In the 1940s, following Yorke's lead, Joseph Sickler, a local newspaper correspondent, set out to photograph many of the same old buildings. Like Yorke, Sickler was most interested in the earliest buildings, but primarily in terms of their historical associations. Sickler extended Yorke's earlier work by including several buildings that Yorke had overlooked, writing short histories of the buildings he photographed, attempting to identify local brick masons and link them to individual dwellings, and compiling his largely antiquarian text and images into a slender book that is well known today by almost all the local history-minded folks as "Sickler" or "Sickler's book." Sickler's book, like Yorke's photograph album, provided an important record of South Jersey's historic architecture and today remains a useful starting point for anyone who wants to learn more about the region's early buildings. Still, Sickler was especially interested in houses with ornamental brickwork, and he thus concentrated on photographing the area's showiest pattern-end dwellings. And Sickler, like Yorke, photographed not only buildings that had been preserved by earlier generations, but also those that were deemed historically significant in his time.[4]

For Yorke, Shourds, Sickler, and others who followed in this documentary tradition, the old brick houses of Salem County functioned as ancestral maps—interconnected networks of artifact, location, and history; linked places of significance that encapsulated much of what was important—and memorable—about the local past.[5] Yet the ancestral maps that were established in the 1700s and reinvented through subsequent centuries have not simply documented the history of the local architectural landscape; they have shaped that landscape through subsequent iterations, not only by historicizing a particular group of buildings and emphasizing their linkage to founding families, but by allowing the dwellings of this elite contingent to symbolize by implication the history of the entire Salem County region. As Joseph Sickler gushed in his introductory essay, "here exemplified in these houses, is the living memorial of a great past, the living history of the oldest English settlement on the Delaware River. For their history is the history of Salem."[6]

But is it? What *do* the buildings tell us? In what way did they relate to other buildings in the area, how were they initially conceived and subsequently altered and occupied, and how did they in turn occupy the landscape? How is the history of the region embedded in this brick architecture? And, when we view these buildings that have been so thoroughly documented, monumentalized, and preserved for the last few centuries, and that are so much a part of our present-day conception of southwestern New Jersey's past, what is it we are actually seeing? The ornamented brick houses referenced by Yorke, Sickler, and Shourds are well-known to architectural historians because of their spectacular patterned, initialed, and dated brickwork and their remarkable regional concentration. And, to most residents of South Jersey today, these houses constitute some of the most visible and articulate reminders of the region's past. Yet these buildings housed a landed and influential minority when they were first built. The individuals who constructed these dwellings in the eighteenth century were typically among the wealthiest and most prominent citizens in the area—the region's predominantly Quaker elite. But because these houses survive today in numbers that mask their original rarity, they represent but one aspect of a far more complex historical and cultural landscape. And because the landscape contexts that once helped to define these early buildings had already, even by the time Yorke and Shourds were working, experienced several overlays of change, the original meanings of these buildings, like the reoriented William Hall house that Thomas Yorke photographed in 1888, had already begun to be clouded by subsequent layers of interpretation.

At the close of the eighteenth century, English and European settlement had shaped the South Jersey landscape for more than a century. Several generations had occupied many of the earlier, heavily documented buildings that were so prominently featured in the work of Yorke and Shourds; and heightened rebuilding activity, which had become especially widespread during the late eighteenth and early nineteenth centuries, had effected significant and far-reaching changes to the aging colonial landscape. By the time Yorke photographed this landscape in the 1880s, it had been so dramatically filtered through the early national period that not a single early eighteenth-century house stood unaltered, and some had experienced more than one rebuilding episode. Moreover, these buildings had been transformed during this critical period in very particular ways. These widespread architectural transformations, which preserved and heightened symbolic content while simultaneously extending and rearranging physical space, had the net effect of redefining the early colonial landscape. This newly redefined "colonial" landscape now embodied goals of preservation as well as alteration, of respect for tradition as well as acceptance of the need for change, of conscious stewardship of a particular vision of the past as well as adaptation for the present. Thus, in this long-settled and "ancient" South Jersey landscape, architecture, even in the early stages of nation-building, had already acquired the aura of a "significant" past, and the "early colonial" landscape that Yorke immortalized was just as much an early national landscape as it was a product of his own late nineteenth-century vision. That vision, which clearly originated in the intentionally monumental quality of the buildings themselves, was refracted through these subsequent alterations to—and reinventions of—the ancestral landscape.[7] And that vision of the past that informed these earlier transformations continues to influence our reaction to the landscape today.

Much of this early national landscape has disappeared. While only a few buildings from this period survive today, documents such as tax lists, inventories, and period accounts allow us to reconstitute a more complete and nuanced view of the early South Jersey built environment. The landscape that emerges from a sustained examination of all of these sources differs significantly from the image that prevails through so many earlier interpretations, and it suggests something of the power and reach of the historicizing process. But first, before we examine the South Jersey landscape in greater detail, it would be helpful to consider some of the factors that influence the complex processes of landscape formation.

Landscape is simultaneously artifact and place. The cultural landscape, probably the largest object of material culture, is the human-modified terrain that forms, in James Deetz's terms, the "connective tissue that gives houses and communities their proper context."[8] At the same time, landscape is also a site, serving to ground historical events and providing a stage for human action. Enormous in scope and profoundly spatial, it thus gains much of its definition not only from its sheer compass, but also by comparison with other human landscapes. But while this very expansiveness suggests a certain stasis, landscapes are continually in flux. The landscape is constantly being re-edited, and earlier landscape expressions are regularly overlaid with those of subsequent generations. Because the cultural landscape demonstrates its temporal dimensions more overtly than many other artifacts, it thus constitutes a kind of human autobiography—the physical manifestation of multiple human actions over an extended period of time.[9] As a kind of material inscription of the human past, then, landscape can also function as a repository for public memory, a touchstone to link the present with the past. Yet, although cultural landscapes embed many past material expressions, the historical layers recorded within them are not always orderly or neatly stratified, nor are they totally representative. Depending on the impact of values and choices through time, some past layers are completely absent, some are pushed forward, and some are more subdued than others. Thus, historical process in the landscape is often collapsed, much like the collapsing of three dimensions into the two-dimensionality of the map. And, as J. B. Harley reminds us, authenticity in maps is relative. As cartographic representations of the landscape, maps always embody values and are neither true nor false. In maps, which are selective in their content as well as in their signs and modes of representation, "the domain of expected detail has been throttled down to a certain class of objects."[10] This "throttling down" redefines the landscape in terms of prevailing cultural values as well as the dominant structures of power. Once created, maps acquire their own authority, which may prove difficult to dislodge. A map, then, is not just a navigational tool; it is simultaneously a way of knowing and articulating the human world which also shapes that world.[11] Like traditional maps, conceptual cartographies that are played out in material terms on the landscape operate in similar fashion. As built features are edited away through time, details are throttled down, and cultural expressions are elided, the cultural landscape is continually changed and the canon of the ancestral map is perpetually reinvented according to prevailing values. Thus, the domestic monuments created in the eighteenth century by southwestern New Jersey's elite landowners are artifacts of past structures of power as much as they are artifacts of the lived and experienced past.

In his study of elite historic landscapes in upstate New York, geographer Peter Hugill advanced the concept of the extended landscape gesture. Gestures—the social constructs or communicative "statements" that people make both within the boundaries of their own social group and in response and relation to other social groups—include not only performed behaviors, such as actions and spoken language, but also material inscriptions, such as cosmetic appearance, clothing styles, place of residence, objects like furnishings and houses, and landscapes. Hugill found that elite groups typically establish and maintain social rank by employing sets of gestures, both material and nonmaterial, that other groups find difficult to replicate—gestures that often coalesce around notions of genealogy, history, and landscape. Still, gestures are constantly being reorganized. Landscape gestures are particularly meaningful statements because they are purposeful material creations that endure for long periods and eventually build up over time.[12] As landscape gestures or ancestral maps extend through time, exerting an impact upon subsequent habitations, they also acquire renewed meaning as they are reappropriated, reformulated, and remapped.

This idea that landscapes owe much of their present and historical appearance to such a lasting imprint of earlier populations is not a new one. Cultural geographers have long maintained that landscapes tend to become imprinted with the social and cultural characteristics of their first effective populations. When territories experience settlement, the argument goes, or when invaders displace or overtake an earlier population, this first viable, self-perpetuating society places its own stamp on the land so forcefully and enduringly that its impact remains considerable even after successive waves of habitation. No matter how small the initial group of settlers is, this first effective settlement produces a lasting impact on a region, and typically means much more to its cultural geography than the contributions of massive numbers of immigrants a few decades later.[13] The marks left by these initial settlements can remain important for decades and even centuries.

In South Jersey, the landscape has followed such a course. Geographer Peter Wacker maintains that much of the entire New Jersey landscape owes its present and historical appearance not only to the imprint of its earliest effective settlements but also to the decisions that followed directly in their wake.

In other words, these traces of past landscapes have served as critical reference points for subsequent changes, as key components of many ensuing physical—and mental—landscapes.[14]

Although the historical significance of the first effective settlers in southwestern New Jersey—the English Quakers—is clear, the reasons behind some aspects of their physical imprint are not entirely transparent. While patterned brick architecture was present in several locations throughout the Mid-Atlantic, the distinctive regional concentration of patterned brick architecture in southwestern New Jersey is particularly noteworthy. Debate about the reason for that concentration continues. Some historians have seized upon the fact that many of the most prominent houses were owned by an entrenched Quaker elite. For example, Michael Chiarappa has maintained that southwestern New Jersey's patterned brick dwellings were a direct reflection of prevailing eighteenth-century sectarian power structures, and that, as demographic and social changes began to erode their established power base during the middle and late eighteenth century, Quakers in southwestern New Jersey may have utilized the built environment to monumentalize their imprint on the landscape. With their prominent brick houses and meetinghouses, South Jersey Quakers thus exhibited an especially dynastic form of geographic consciousness—one that constituted an aggressive expression of what some scholars have termed Quaker tribal authority.[15] Often patterned, conspicuously initialed, and carefully placed, these brick houses were the aesthetic correlatives of Quaker meetinghouses. With their dated gables and careful brickwork, dwellings and meetinghouses resembled each other in an ongoing architectural dialogue; but because the distinctions between the two architectural forms were intentionally blurred, Quaker geographic authority in the region derived not just from the dwelling house or the meetinghouse alone but from the symbolic power embedded within the relationship between the two. Quaker pattern brickwork thus lent regional Quakerism a social and cultural power that extended beyond the established institutional framework and constituted a material and territorial response to Quakerism's dwindling presence in mid- and late-eighteenth-century southwestern New Jersey.[16]

Sectarian tribal authority may not be the only explanation for such a lasting landscape imprint. Socioeconomic status could have played an even larger role. While most of the inhabitants and builders of the conspicuous brick houses and meetinghouses were certainly Quakers, they were also, like the "mansion people" of the Connecticut River valley, among the wealthiest and most prominent individuals in their communities.[17] Moreover, because these buildings often carried the initials of both female and male marriage partners, these objects monumentalized not just the work of individuals, but the imprint of entire families.

This kind of socioeconomic status could also translate into political power. In his work on the social origins of colonial New Jersey assemblymen, Thomas Purvis has termed the eighteenth century "New Jersey's age of oligarchy." A small number of elite families managed to consolidate political power fairly effectively throughout the state, and this age of oligarchy began to erode only when the old leading families started to lose power during the Revolution. Wealth and social status, Purvis found, were far more important determinants of political influence than was religious affiliation.[18] Thus, like the elites that Peter Hugill studied in upstate New York, the material inscription mapped by the South Jersey Quakers may have been more of a product of an elite world view than that of a sectarian one alone. While this group's landscape imprint may relate to an aggressive geographic consciousness that was exemplified in part by their efforts at regularized meetinghouse placement (see Figure 4.2), it relates just as much, if not more, to the fact that they were an elite and wealthy group with the socioeconomic power in hand to structure—or leave their "gesture" or "ancestral map" upon—the landscape.[19] That this shaping of the eighteenth-century environment has prevailed through ensuing landscape iterations suggests as much about subsequent—and present—power structures and value systems as it does about those of the mid-eighteenth century. In southwestern New Jersey, as Thomas Yorke's photographs attest, this map of the ancestral landscape has experienced periodic readjustments through time, yet most of these landscape redefinitions have related directly to prevailing structures of power.[20] Some of the most enduring of these landscape readjustments occurred in the region's second century of settlement, almost one hundred years before Yorke and his contemporaries spurred renewed interest in the ancestral landscape. The example of Mannington Township in its Salem County environs provides a case in point.

Mannington Township adjoins the town of Salem, which was one of the earliest settlements in the Delaware Valley (Figure 4.3). Although the township does not actually abut the Delaware River, its flat, low-lying landscape is both

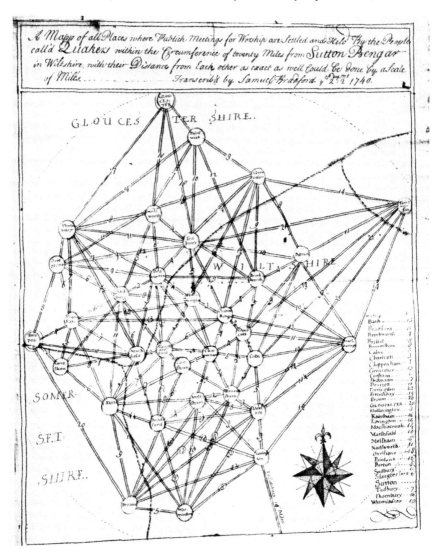

FIG. 4.2. Samuel Bradford's map showing regularized placement of Quaker meetings within twenty miles of Sutton Berger, 1740. Courtesy Bevan Naish Library, Woodbrooke Quaker Study Centre, Birmingham, England.

delimited by and laced with water. This is a landscape of brackish marshes, flat and gently rolling pastures, patches of forest, and numerous creeks and streams.

The region's multiple accessible waterways and fertile soil encouraged early, intensive, and durable settlement. By the time of the Revolution, Salem

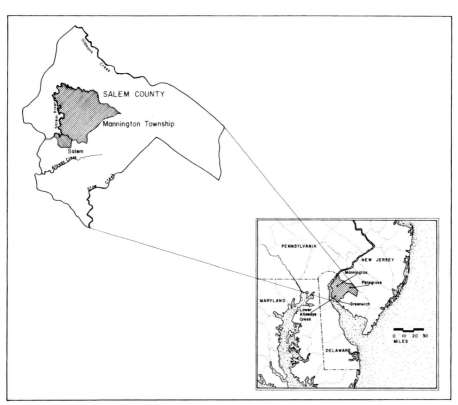

FIG. 4.3. Mannington Township, Salem County, New Jersey

County had been settled for more than a century, first by Fenno-Swedish and Dutch immigrants, in the 1640s, and later by a sizable group of English Quakers led by John Fenwick in 1675. Subsequent in-migrations of Quakers, mostly from northwestern England, Middlesex, and London, strengthened this early settlement, and by the second decade of the eighteenth century, the English presence dominated.[21]

In his promotional pitch to potential settlers in 1675, John Fenwick had described Salem County as a "terrestrial 'Canaan'" where "the land floweth with Milk and Honey."[22] Ten years later, surveyor Thomas Budd used slightly less effusive but no less positive terms when he described the land along the Delaware Bay and River as "big rich fat marsh land."[23] Because the region's soil was so fertile and because water transportation made distant markets so accessible, agriculture dominated the region's economy. This was a region of dispersed farmsteads in the eighteenth and early nineteenth centuries; wheat, corn, rye, livestock, and forest products drove the economy. Salem also became an active port in the eighteenth century, shipping grain, lumber, and agricultural products. And when the Chesapeake and Delaware Canal opened across the river from Salem in 1829, a new Chesapeake market was brought within reach.

Like the rest of Salem County, Mannington's geographic proximity to Philadelphia and other markets proved advantageous for agricultural development, as not only the Delaware River but also the many smaller creeks and inlets that laced the area constituted important transportation arteries for farmers shipping their crops. Ready access to a landing proved to be a strong selling point in period real estate advertisements. For example, an advertisement for a farm in adjacent Penn's Grove called the property "very convenient for Philadelphia Market, as Flats and Shallops are almost every Day going by the Premises, where there is an exceeding good Landing, and Shallops often take in their Loading there."[24] As the advertisement for one Mannington farm noted, Salem itself boasted a public landing "from whence any produce may be sent to Philadelphia market for a trifling expence."[25] The farms with the best transportation access were usually the most desirable locations. The area's water orientation also reinforced the transportation of bricks for building purposes and allowed bricks to be shipped expeditiously, usually via flat-bottomed scow, to other locations, which tended to be water-oriented as well.[26]

This water network encouraged frequent interaction between the inhabitants of southwestern New Jersey and the rest of the Delaware Valley and stimulated a continual social and cultural exchange. Numerous architectural commonalities between eighteenth-century buildings on both banks of the Delaware River attest to the early establishment of this easy cross-river dialogue. Period accounts, too, suggest the frequency with which river travel routinely occurred. In 1773, the crossing between Port Penn, Delaware, and the Cohansey Creek in South Jersey took about thirty-five minutes in a small wherry.[27]

Still, the southwestern New Jersey landscape remained just as locally distinct as other parts of the Delaware Valley. Farming, of course, was important, but not all farmers owned the land they worked. Tax lists from 1773 through 1797 reveal a high degree of landlessness throughout the township.[28] In 1773, more than half of the population owned no land; although landlessness declined gradually through the end of the century, in 1797 nearly half of the population still owned no land. Compared to Pittsgrove, a nearby Salem

County township located just to the east and farther inland, landlessness in Mannington was somewhat more common, perhaps because in Mannington fewer property owners controlled significantly larger tracts of improved land and unimproved land accounted for a very small percentage of the total land in the township.[29] Not surprisingly, this rate of landlessness in Mannington corresponded closely with levels of overall wealth. In 1797, the top one-fifth of the population controlled more than one-half of all of the wealth in the township, while the landless 40 percent held less than one-tenth. The owners and occupants of Mannington's few brick houses were among these wealthiest landowners (see Tables 16–18 in Appendix).

Accordingly, tenancy was common. Compared to more northerly parts of New Jersey, Mannington's tenancy rate, which stood at 44 precent in 1798, was unusually high and corresponded closely with its rate of landlessness. Tenancy in several other parts of Salem County was similarly high. But while landlessness was a widespread condition throughout the area, tenancy did not necessarily equate to dramatically inferior living conditions. As historians have demonstrated, tenancy in many parts of the Delaware Valley was a viable economic strategy that could benefit tenants as well as landholders. Functioning as a sort of socioeconomic ladder, tenancy offered renters a chance at upward mobility while it enabled wealthy landowners to maintain and improve the productivity of their land.[30] In Mannington and neighboring Lower Alloways Creek Townships, differences between the way owners and tenants lived appear subtle despite significant economic cleavages. Owner-occupied houses in Mannington were usually worth more than tenanted dwellings but were only slightly larger and only slightly more likely to be built of costlier brick.[31] In addition, by the end of the eighteenth century, tenants occupied several of Mannington's few brick houses. Even though these buildings had almost all been the "best" houses when they were originally built, by the late eighteenth and early nineteenth centuries most were at least fifty to seventy-five years old and represented aging housing stock.

Local distinctions also emerged in the way landowners used the land itself. Land parcels in Mannington, like those in much of South Jersey, tended to be significantly larger than those farther to the north, averaging around 130 acres (see Table 19 in Appendix). Compared to some other parts of the Delaware Valley, such as the fertile Lancaster Plain and parts of Chester County, Pennsylvania, average landholdings throughout southwestern New Jersey were also somewhat larger.[32] In his studies of land use in early New

Jersey, Peter Wacker found, not a degree of regional sameness, but a distinct and localized mosaic of differences in landholding patterns across the state. Wacker discovered that in the northern parts of New Jersey—those areas, such as Essex County, settled primarily by New Englanders—farmsteads were small, averaging 72 acres in 1780, but that landholdings in the more southerly, English Quaker–settled areas, such as Salem County, were more than twice as large, averaging about 167 acres. Wacker also found that a high percentage of the farms in much of southwestern New Jersey were exceptionally large: between 31 percent and 50 percent of all farms in Mannington, Upper Alloways Creek, Pittsgrove, and Woolwich Townships and more than half of all farms in Elsinboro Township contained more than 200 acres. The size of these large land parcels generally corresponded to the region's high tenancy rate, for the wealthy landowners on this fertile inner coastal plain had established a system of large landholdings worked by a landless white labor force.[33] Thus, in Mannington as well as in nearby parts of Salem and Cumberland Counties, an elite landholding class continued to maintain its hold on the rural landscape through the end of the eighteenth century by sustaining the high rate of landlessness and preserving the pattern of large land parcels which had prevailed from the earliest days of settlement.

By 1798, most of Mannington's agricultural landscape, like much of the rest of southwestern New Jersey, was improved.[34] Improved land (land that had been surveyed from the collective proprietary domain and had experienced any improvements at all) was taxed. Improved land also translated into productive crop yields and higher land values overall. In the early 1700s, livestock had grazed on much of the unimproved land in Salem County, and uncultivated forests blanketed the remainder; but after midcentury, the landscape gradually changed. By 1784, Salem County was more than three-fourths improved, and at the close of the eighteenth century, land in Mannington itself was over 96 percent improved—a dramatic contrast with other areas in the Delaware Valley region, such as parts of Sussex County, Delaware. Perhaps because Mannington included some of the best soil in the county, it outstripped many other areas in land improvement (see Table 20 in Appendix). Still, unimproved woodland, much of which was used to supply staves and cordwood that was transported by flatboat to the nearby Philadelphia market, lingered in parts of the inner coastal plain as late as 1780. Improved land was always more highly valued than unimproved acreage, and valuations for improved land in Salem County were among the highest in New Jersey.[35]

An agricultural economy that emphasized grain, livestock, and dairy products prevailed throughout the region. On his farm in Cohansie (also spelled "Cohansey," in what is now Cumberland County), Philip Fithian harvested salt hay and planted grain crops such as rye, wheat, and corn. Richard Wood, who lived in nearby Greenwich, planted timothy, corn, oats, and flax, harvested salt hay, and raised cattle for beef and cheese.[36] Toward the end of the eighteenth century, real estate advertisements for southwestern New Jersey often highlighted the fertile soil and advantageous locations that permitted "garden truck" to be raised "in great plenty and brought to Philadelphia market by water."[37] Advertising copy usually emphasized such features as proximity to water, land quality and fertility, the abundance of timber for cordwood and fencing, and the types of pasturage and crops that the land would support. Such advertisements also usually listed the types of improvements that might be possible or the condition of any improvements that had already increased the value of the land.

In Salem and Cumberland Counties, especially, improvements to land more often than not included tide banks or dikes, earthen embankments built to control drainage on land abutting waterways. One Mannington plantation located "within three miles of Salem town" typifies this late eighteenth-century tidewater landscape. This farm contained "about 600 acres; 240 whereof are new banked meadow ground" abutting both Mannington and Salem Creeks and including about 60 acres of "old drained meadow" that "would suit for hemp as well as grass."[38] From an early date, the abundance of water and the overall marshiness of this fertile but low-lying tidewater landscape had demanded a number of specialized agricultural solutions such as the "banked meadow ground" or reclaimed marshland encompassed by this Mannington plantation. The reclamation of the marshes constituted one of the most significant and lasting changes to the southwestern New Jersey landscape. As early as the late seventeenth century, marsh reclamation was practiced in southwestern New Jersey as well as in parts of Delaware on the opposite bank of the Delaware River. The tradition of diking land came from England and Holland and tended to take hold in the New World wherever large numbers of English or Dutch settled in marshy coastal areas.[39] Although historians credit the Dutch with introducing the technique of diking to the lower Delaware Valley, the English Quakers seized upon the technique and implemented it most significantly (see Figures 4.4a and 4.4b).[40] Banked meadows constituted an important aspect of the local agricultural economy.

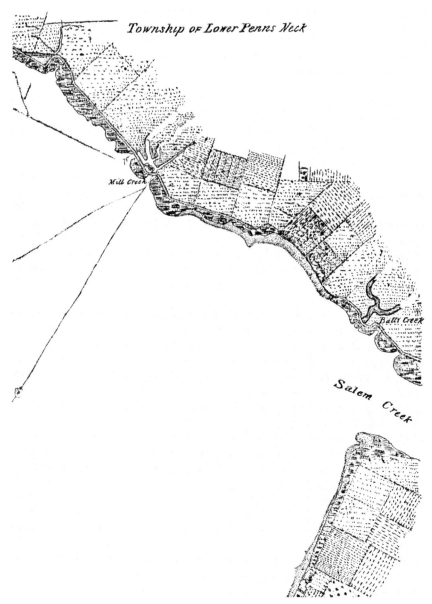

FIG. 4.4A. The tide banks on this 1841 coastal survey map show as dashed parallel lines running along the coast. U.S. Coast Survey, A. D. Bache, superintendent, "Part of Delaware River from Peapatch Island to Reedy Island," by F. H. Gerdes, assistant, U.S.G.S., 1841. National Archives.

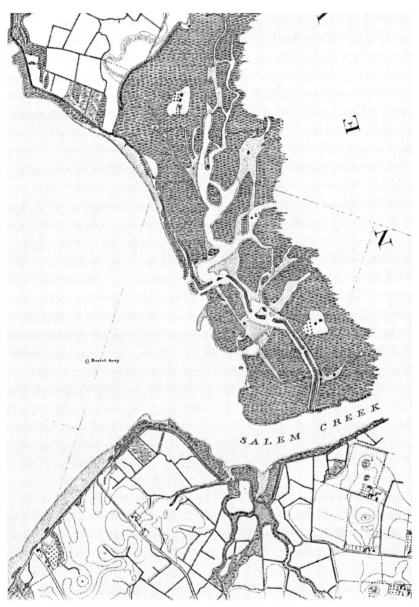

FIG. 4.4B. This 1882 coastal survey map also shows tide banks. U.S. Coast and Geodetic Survey, J. E. Hilgard, superintendent, "The New Jersey Shore of Delaware River from Kelly's Point to Elsinborough Point," by R. M. Bache, assistant, July, August, and September 1882. National Archives.

Diking salt and freshwater marshes enhanced the production of hay, which, as Peter Kalm noted, had often been in such short supply in the mid-eighteenth century that winter travelers passing through New Jersey, Pennsylvania, and Maryland sometimes had trouble feeding their horses.[41] Salem and Cumberland County farmers used an arrangement of sluices, gates, and dikes or banks to block tidal waters and permit the marshes and meadows to drain at low tide.

Reclaimed tidewater meadows were often used to pasture grazing livestock. Unfortunately, cattle would sometimes stray into the alluvial mud of the drainage ditches and become mired in the muck. One March day Philip Fithian wrote, "To day we had a creature mired but likely to recover." The following month, another animal was not so lucky: "Sunday we had a cow mired, we looked for her on Sunday towards evening but did not find her. Monday we all went after her, about 11 I found the poor creature, lying dead in the mire." Similarly, in May of 1812, Richard Wood noted "having a cow got in the ditch we had to go and pull her out."[42]

On a few occasions the high water table and general marshiness of the low-lying landscape caused other problems. During one particularly wet spring, Philip Fithian remarked, "The hollows are filled with water and overflows running across our street in two places constantly by reason of the springs rising to the surface of the Earth." That year, the water was especially high, and he wrote, "the cellars of numbers are almost filled with water and afford enough for use of the family."[43]

Real estate advertisements from the middle and late eighteenth century regularly cited the amount of improved land "within bank" as well as the newness, length, and overall condition of the banks. Advertisements typically noted such improvements as the "30 acres of good dry meadow, made out of tide marsh" on one Mannington farm, or the perceived potential of land such as the 200 acres of land on the Delaware River "on which is and can be made 60 acres of excellent bank and swamp meadow" or another nearby 120-acre tract "of which 25 acres of excellent swamp meadow may be and is made."[44] One advertisement, which bragged that the land in question was some of the "best land in that part of the Jerseys," boasted, "there are about 40 acres of drained meadow, which require but a short bank" as well as around 200 acres of "exceeding good high marsh, firm enough to bear a loaded team over any part of it."[45]

Banked meadow thus constituted a distinct real estate selling point. Still, building tide banks was a costly, time-consuming, and labor-intensive affair. It typically involved staking out the site for the intended bank and digging a trench parallel to but set back a distance from the margin of the river or creek which bounded the marsh. This setback allowed for the overflow of swells and high tides in stormy weather. Mud was then removed from the trench, transported across the miry marsh via wheelbarrows rolled atop planks, and piled load upon load onto the bank. The size of the finished bank depended on geography and circumstance, but according to Robert Gibbon Johnson's 1826 treatise on reclaiming marshland, "a bank of twelve feet base by six feet in height, would be sufficient, but," he cautioned, "I would observe, that I have never seen a bank too large." Johnson, a major southwestern New Jersey landowner, was also a prominent local agricultural reformer and member of the Philadelphia Agricultural Society. Like marsh reclamation schemes espoused by wealthy slaveowning plantation owners in the South Carolina low country, Johnson's method required mountains of capital, days of work, and large gangs of laborers. To build the bank properly, according to Johnson, these laborers "should be apportioned to the different parts of the intended bank in separate gangs, as the work will then be carried on to the best advantage, not only to themselves, but their employers." A gang of laborers was "composed of five hands, two to cut and load the wheelbarrows with mud, two to wheel it to the bank, and one to be there stationary as a packer" whose job was to act as a sort of mud mason, "attend[ing] strictly to the carrying up the bank, in its true proportion and proper height."[46]

Once constructed, tide banks also needed constant repair, as weather, tides, and burrowing muskrats could all wreak havoc.[47] Philip Fithian devoted considerable time and energy each year to maintaining the banks and repairing the sluices and gates which kept his Cumberland County farmland productive and allowed the drainage ditches in his meadows to empty into the Cohansey River at low tide. Fithian periodically repaired the banks by applying layers of mud carted in barrows or canoes.[48] Richard Wood, who regularly harvested and scowed salt hay from his banked Greenwich meadows nearby, also attended to "boging my meadows & heaping the sods" as well as repairing occasional breached banks on his meadows abutting Stow Creek. Still, Wood often complained of the difficulty of haying during particularly wet seasons, as in August of 1824, when he remarked, "I never had so distressed a time in getting salt hay on account of rain & tides."[49] Similarly, Robert Gibbon Johnson, the published proponent of marsh reclamation, who also held much of Mannington's acreage and owned some of the largest land tracts in Salem County, regularly referred to repairing the sluices and banks that riddled his reclaimed marshland.[50]

Robert Gibbon Johnson and Richard Wood typified the elite farmers whose land improvements had such a far-reaching effect on the area. Throughout the South Jersey region, banked meadows were almost always associated with wealthy landowners like Johnson and Wood in the late eighteenth and early nineteenth centuries. Although one reason for this may simply have been the high cost and labor intensiveness of tide bank construction, Peter Wacker has suggested that the practice of banking meadows in New Jersey may have been associated with an agricultural elite for additional reasons. Wacker contends that the concentration of banked meadows in lower New Jersey and their general absence throughout the rest of the state is particularly significant because salt and freshwater marshes in other parts of the state that *could* have been diked ended up being treated very differently. Marshland, which could be reclaimed by digging a series of ditches and drains as well as by building tide banks, tended to be banked only in southwestern New Jersey. In areas that were settled primarily by New Englanders, however, such as some of the more northerly parts of New Jersey, meadows tended to be ditched and fenced, and bridges were sometimes built over the ditches, but such meadows were rarely diked. Wacker has suggested that land improvements such as banking may have been characteristic of elite "farmers who belonged to agricultural societies, read the agricultural literature of the day, and had the capital to invest in reclamation schemes."[51] Certainly, the wealthy Quakers in the region had been linked by some observers to the practice of banking tidal meadows even as early as the late eighteenth century; John Rutherford noted in 1776 that the Quakers, "a frugal and industrious People" had "banked off many valuable meadows."[52]

However, even the wealthiest landowners found the cost of marsh reclamation to be high. Because banking and draining projects tended to be so expensive and because land drainage affected surrounding properties, marsh landowners in southwestern New Jersey often banded together in local marsh or meadow companies to pool reclamation costs. Marsh companies were private organizations that levied their own taxes, usually proportionally based upon the amount of marshland owned, in order to finance specific marsh improvements. Although many marsh companies on both sides of the Delaware River began and flourished in the nineteenth century, the era of widespread agricultural reform, some had formed even as early as the late seventeenth century. Salem County, which boasted the largest amount of reclaimed marshland in the southwestern New Jersey area, was also home to the oldest and largest number of meadow companies.[53] The formation of meadow companies here was supported and encouraged by a 1788 law enacted by the New Jersey state legislature which assisted owners and renters of tidal marshes either to improve their property through reclamation or to maintain previously reclaimed land. This law, which extended a similar law passed by the colonial legislature in the early eighteenth century permitting area farmers to incorporate as meadow companies, required such organizations to meet annually, discuss business, and elect officers who would be responsible for collecting funds from other company members and overseeing the construction and maintenance of the banks.[54]

One of the earliest and most long-lived meadow companies in southwestern New Jersey was the Mannington Meadow Company, an organization that typifies the way these marsh companies began, grew, and changed over time.[55] The Mannington Meadow Company formed in 1756 with a total of seven members, all of whom owned meadow and marshland abutting Mannington Creek and many of whom were some of the wealthiest landowners in the township. In fact, the company's first recorded meeting took place at the home of William Hall—probably the same brick house that Thomas Yorke so carefully photographed more than a century later. The earliest company activities involved choosing managers, assessors, and a collector and settling debts for specific improvements such as payment for "diging the hole and Laying in half the Sluse" or "mending the dam" and "giting the timber for stoping krick [creek]." Through the rest of the 1700s and into the first three decades of the nineteenth century, the company members met and recorded their activities annually, each year choosing their officers, collectors, and assessors and deciding when to open and close sluice gates, where to cut ditches, build tide banks, and lay sluices, and how to pay for them.

This small band of marsh owners also became more bureaucratic and businesslike over time. In 1759 they were referring to themselves as the "owners & possessors [renters] of the Drained Marshes & Meadows on Mannington

Creek," and by 1802, the members called themselves the Mannington Marsh Company and operated under a new "act for maintaining the banks and Sluices, and draining the meadows on Mannington Creek." By 1823, there were thirty-three members representing a total of 607 ½ acres of marshland. This same meadow company that began as a small band of elite property owners continued to control the landscape by representing the interests of the wealthier Mannington landowners in the late 1700s and through the first decades of the nineteenth century (see Table 21 in Appendix). As with so many other contemporary agricultural reform movements, the capital-intensive reclamation of the southwestern New Jersey marshes, which effected a major regional landscape change that continued to affect subsequent generations, was both begun and propelled in large part by a regional rural elite.[56]

Communities such as this one in southwestern New Jersey in which elites control water resources have been termed "hydraulic societies" by several historians. A hydraulic society is based on the notion that control over water is a way of consolidating power and giving some people power over others. In his study of water and the American West, Donald Worster defines a hydraulic society within a mostly arid context. He sees the modern hydraulic society of the American West as an "increasingly . . . coercive, monolithic, and hierarchical system, ruled by a power elite based on the ownership of capital and expertise."[57] In other words, by remaking nature, elites assert their political, social, and cultural dominance. Such hydraulic societies tend to follow one of three broad patterns. In the simplest and least coercive of these patterns—what Worster terms the "local subsistence mode"—water control is effected by means of small-scale and temporary structures that exert minimal interference upon natural water flows. Power at this stage tends to coalesce along family and kinship lines because hierarchy and discrimination are not really necessary. This localized system begins to erode, however, when individuals unite into broader regional authorities in order to control water. When this happens, when local self-management at the individual level no longer suffices, "a more efficient utilization of rivers . . . and a more elaborate legal framework to resolve conflicting interests" begin to occur. With this second mode of control—the "agrarian state mode"—the more intensive and ambitious water control schemes that develop simultaneously produce a managerial elite. The emergence of such an elite, in turn, is not only accompanied by the loss of autonomy to an entrenched authority but also often involves some sort of bureaucratic organization charged with creating and administering the water system. The third and most coercive mode of water control—the "capitalist state mode"—involves a modern hydraulic society in which water control is carried out by mammoth and comprehensive systems like those of the modern American West.[58]

Worster applied this notion of a hydraulic society to the arid American West, but others have utilized the same concept to explain societies that emerged in far wetter environments, such as those of southwestern New Jersey. More recently, Mart Stewart has argued that wealthy rice planters in the Georgia low country manipulated the landscape as well as resources and human labor to establish a productive environment for growing wet-country rice. Like the elite property owners along the southwestern New Jersey coast, Georgia rice plantation owners engineered the low country environment for profit as well as to enhance their own reputations. They redefined that landscape by channeling the tides and controlling the flow of water with miles of embankments built by gangs of enslaved African Americans. Because banked plantations required a major investment in labor, the first such plantations in Georgia did not appear until the 1760s, when wealthy slaveholding planters began to migrate into the region from South Carolina. This small and powerful elite continued to acquire land, wealth, and slaves and ultimately reshaped the post-Revolutionary low country landscape with their banked plantations.[59]

Embanking a Georgia low country plantation was just as labor intensive as building tide banks was farther north. In Georgia, slaves transported massive quantities of muck and earth to create the elaborate system of perimeter banks, cross banks, and the gridwork of trunks, drains, ditches, and flood gates that characterized most wet rice plantations. The "hydraulic machine" that emerged in the Georgia low country in the eighteenth century and early nineteenth century depended not only on large amounts of labor but upon strict discipline and firm managerial authority, and "the domination of nature and the domination of one group of humans by another evolved mutually."[60] Furthermore, Georgia low country planters understood intimately the power they gained from controlling the tides. In fact, to some planters at least, the elaborate system of tide banks may have seemed more monumental than their own houses. Roswell King, a wealthy planter, wrote tellingly, "the moving of dirt [to build tide banks] appears more permanent than creeting [creating] Buildings."[61]

While the banked tide farms of southwestern New Jersey did not depend upon the same level of overt social hierarchy that characterized the slave plantation economy farther south, they required equally massive amounts of labor and capital to build and sustain. But although the type of dominance these New Jersey elites exercised may have been far less obvious than that of low country planters, it was, in some ways perhaps, more pervasive. Even though Mannington's elite tide bank farmers lacked the gangs of chattel labor that so dramatically reshaped the Georgia low country, they were able to extend their influence over the landscape in a different way, by banding together in collectives like the Mannington Meadow Company. In this fashion, these landowners resembled Donald Worster's second or "agrarian state" model of the hydraulic society: they had outgrown small-scale localized water control methods and had evolved to the point where they needed to create a bureaucratic organization to administer their water system. They also monumentalized their status through this process and confirmed their position as a managerial, as well as an economic and social, elite. Thus, if the labor-intensive "moving of dirt" to create banked plantations symbolized lasting power to Georgia low country planters, the equally capital- and labor-intensive reshaping of the New Jersey landscape must have also suggested, at least in some measure, the widespread dominance of that region's rural elite. These tide banks overseen by marsh and meadow companies constituted only semipermanent improvements to the agricultural environment, but they formed lasting physical symbols of an eighteenth-century elite landscape vision. Marsh companies merely solidified and institutionalized this elite vision so that it could be carried forward.

Many of the more prominent meadow company members were, like Robert Gibbon Johnson, who was a member of the Mannington Meadow Company and a major Mannington landowner as well as a prominent agricultural reformer, among the wealthiest landowners. It was they who inhabited the monumental brick houses like those that Thomas Yorke photographed. Brick dwellings like Johnson's two-story house represented only a small fraction of the southwestern New Jersey housing stock in the late eighteenth and early nineteenth centuries. Brick was a highly valued building material and, as Michael Chiarappa pointed out, a highly transformed one. Because of the many steps involved in its manufacture, brick was one of the most "artificial" of all available building materials.[62] Perhaps because it was such a lasting, transformed, and artificial material, it also became a material that was culturally esteemed and thus became closely linked with so many of the domestic monuments left by southwestern New Jersey's elite.

In Mannington, brick houses were typically more than one-and-one-half times more valuable than log or frame dwellings. Similarly, the land parcels they stood on were usually larger and more costly than others. In 1797, while the average landowner worked about 130 acres, the average owner of a brick house worked a land parcel that was almost two thirds larger and more highly valued. Likewise, owners of brick houses typically kept more livestock. Similar patterns occurred elsewhere in southwestern New Jersey. In nearby Lower Alloways Creek, for example, not only were brick houses usually larger than others in the township, but their owners were much more likely to own other buildings, including a barn and a separate kitchen, and they were much more likely to own rather than rent their house and land. In Pittsgrove, located farther to the east, like patterns prevailed. (See Tables 18–19 and 22–24 in the Appendix.)

If brick dwellings such as those which were monumentalized by Yorke, Shourds, and Sickler and which survive today in such a remarkable concentration made up so small a percentage of the total housing stock, what other types of houses and outbuildings stood on the early national landscape? In Lower Alloways Creek, where only 7 percent of the landowning population owned brick houses, the brick houses were almost twice as large as their frame or log counterparts,; the typical dwelling in that area in 1798 was built of wood and encompassed only about 650 square feet including all floors. On average, this translated into a two-story 16-by-20-foot house consisting of a single room on each floor or a one-story 20-by-32-foot house that was divided, more than likely, into two rooms. More than half of the houses were built of wood, and a miniscule percentage (fewer than 1 percent) were constructed of stone.[63] Only about one in four individuals in Lower Alloways Creek owned a barn or a separately enumerated kitchen, and additional outbuildings such as an occasional one-story wooden stable or shop were extremely rare.[64]

In southwestern New Jersey, the floor plans that those predominantly wooden walls enclosed varied. Local house plan types in the first three-quarters of the eighteenth century ranged from the simplest one-story, one-room plan dwellings to multistoried arrangements involving several rooms to more "modern" closed-Georgian-plan houses with a passage opening onto two or more rooms. Still, a degree of architectural conservatism prevailed. The most widely built forms throughout the region represented a fairly narrow band of architectural options that favored tradition over innovation. Overall, Salem County residents tended to reject closed Georgian forms between 1700 and

1774, opting instead for variations on more traditional hall and hall-parlor arrangements. Additions and changes made to these houses, even in these early years, also materialized in time-honored forms, and they were planned in such a way that the traditional nature of the dwelling was not compromised.[65] Subsequent changes to these houses in the early national period and through the first decades of the nineteenth century generally continued to favor tradition over innovation. By the time Thomas Yorke was photographing southwestern New Jersey's earliest houses, in the 1880s, almost every house had been altered in some way. Most had received one or more additions, and often onto the side rather than the rear. Several had lost original pent roofs or had been raised from one and one half to two stories. Many, like the William Hall house, had been completely reoriented or had had several earlier window and door openings filled in. But when these buildings were altered, they were generally reconfigured with an eye toward preserving the ancestral past. (For photographs and floor plans of the houses discussed below, see the gallery of illustrations at the end of this chapter.)

In the eighteenth century, one of the most common house forms in southwestern New Jersey as well as the entire Delaware Valley region consisted of a single room. Dubbed a hall plan or one-room plan, this simplest and most basic type of house stood only one story high and consisted of a single heated room that incorporated cooking, sleeping, artisanal work, and all other living functions. The front door of a one-room plan dwelling opened directly into the living space, which was typically heated with a single large fireplace on one wall and lit with one or more windows. A ladder stair often led to an unfinished and unheated attic loft that was used for sleeping or storage. More elaborate one-room plan houses, which rose a full two stories in height, are often termed chambered hall plans and included heated rooms on each floor. These second-floor spaces might be subdivided into two rooms, and some included a separate stair passage.[66]

Because one-room dwellings were often enlarged, perhaps as soon as their owners' means permitted, they are not as heavily represented in probate inventories and among surviving dwellings. And because of their size and modular nature, their overall plan also tended to be transformed more thoroughly by additions. Today, since many surviving one-room dwellings have received additions, mostly in a lateral direction, they typically more closely resemble hall-parlor or center passage plans. But close field inspection can reveal the original plan configuration, and the traditional appearance of the original dwelling was often maintained through the changes.

The Kiger house (Figures 4.6 and 4.7), a brick chambered hall dwelling overlooking Salem Creek, typifies the overall size and spatial arrangement of most such dwellings. Yet, because it is built of brick, it hardly typifies the most common one-room houses in Mannington or in Salem County. Like many other monumental brick dwellings in the area, the building overlooks a once-major water access route. The house is a three-bay-wide, single-pile, two-story building with an interior gable chimney and a pent roof on its front (south) elevation. While the facade is built of brick laid in plain Flemish bond and the eastern and northern elevations are laid in a less-ornamental three-course common bond, the western gable, which would have been one of the house's most public sides and the one that would have been seen first on the circuitous approach from Salem inland along the creek, is laid in Flemish bond with ornamental glazed headers. A projecting two-course belt course, also embellished with glazed headers, encircles the entire building. According to Sickler, the house was built by a Richard Robinson in 1720. A brick incised with the initials R.R. and a later date of 1745 is still visible over the front door. By 1744 the house was tenanted by the two Kiger (or Geiger) brothers, who were among the first in a group of Belgian and German workers employed at the nearby Wistarburgh Glass Works. Locally the Kiger house is also known as the Jesuit mission house because it functioned as a church for about twenty years during the period when Catholic services were banned.[67]

The Kiger house was not among the buildings photographed by Thomas Yorke in the 1880s, so we lack his visual record of its late-nineteenth-century appearance. Today the building displays much of its early-eighteenth-century exterior appearance because of the way it has been enlarged. Instead of receiving a lateral addition, the original main block has been extended by means of two end-to-end frame additions off the back (which may have once been the front and more public side) of the house, leaving the ornamental patterned gable untouched.[68]

Measuring nearly 24 feet wide by 18 feet deep, the original house consisted of a single room on the first floor with centrally placed doors on both front (south) and rear (north) elevations. Two windows flanked the front door. A window on the present rear elevation may have offered additional light, but evidence suggests that this window opening was first moved and then subsequently filled in when the frame additions were attached. A single large fireplace on the paneled west wall (Figure 4.8) heated the main room. An enclosed winding stair abuts this fireplace and leads to the second floor, which consists of a large unheated room, a smaller room with a fireplace, and a small unheated room over the stair. Although much of the original paneling and other interior finish has been lost over time, the house was restored in the twentieth century. With a footprint of only 480 square feet, the Kiger house is atypical of the brick houses in Mannington, as it was relatively small and less highly valued than most other brick dwellings in the township.

Throughout southwestern New Jersey, for hall-plan and chambered hall dwellings like the Kiger house, adding a smaller wing to a gable end became the preferred method of enlarging dwellings. The William Nicholson house, which once faced a tributary of Mannington Creek, was constructed circa 1730, for one of Mannington's elite landowners, as a chambered hall brick dwelling with a floor plan resembling that of the Kiger house.[69] The house was first enlarged, probably very shortly after it was built, with a slightly smaller, slightly lower brick kitchen wing attached to its gable.[70] By the time Yorke photographed the house in the 1880s, a frame lean-to abutted the kitchen wing and both brick sections had lost their original pent roofs, but both still stood intact behind a white picket fence (Figure 4.9). As he did with the William Hall house, Yorke photographed what was then the front of the building, for, like the Hall house, the Nicholson house had been reoriented by changes in transportation routes. The original facade now faces a more private landscape of cultivated farm fields rather than a landing on a navigable waterway. In subsequent years, the house was extended even further by means of a two-story frame addition and a one-story log house moved up to the opposite gable, so that now an approaching observer sees the two earliest brick sections as well as these later additions (Figure 4.10).

From the outside, both the placement and the design of the brick kitchen addition carefully emphasized and preserved the ornamental and monumental qualities of Nicholson's original chambered hall dwelling. The chevron pattern of glazed headers beneath the rake of the kitchen roof is carefully matched to a similar pattern adorning the main block, and the kitchen gable is further ornamented by a stepped glazed header belt course. Yorke's photograph also clearly shows a brick cartouche, suitable for a datestone, located near the peak of the main block, just below the raking line of glazed headers.

The roof ridge of the kitchen addition joined this gable just below these significant features, so that they were still clearly visible. The traditional nature of the existing interior space was similarly preserved (Figure 4.11); the kitchen addition simply transformed a hall-plan into another open plan type: a two-room or hall-parlor plan.

Another common early house type in the region began with a somewhat larger footprint, with two rooms on the ground floor instead of one. By far the house plan most frequently mentioned in probate inventories and represented by surviving buildings is a two-room design known as the hall-parlor plan.[71] Hall-parlor dwellings were one room deep and two rooms wide, with their two rooms arranged end-to-end and parallel to the roof ridge. The two first-floor rooms usually consisted of a public hall or kitchen, which was furnished with the main cooking fireplace and a stair or ladder to the loft, and a more private room that was used as a parlor, as a sitting room, or for sleeping. In hall-parlor dwellings as in one-room plan and three-room *flurküchenhaus* plan houses, the front door opened directly into the heated living space. Although hall-parlor houses with unheated parlors were built in the region, most had fireplaces in each room. Also, while some hall-parlor houses were built to be, and remained, one story tall with just attics or lofts over the ground-floor rooms, most of those that show up in eighteenth-century Salem County probate inventories rose to a full two stories and contained at least two additional rooms on the second floor. Most hall-parlor houses in southwestern New Jersey exhibited a relatively uniform size, ranging from about 16 or 17 feet on the gable walls by around 25 to 35 feet across the front.[72]

Several early brick hall-parlor houses survive in Mannington. Unlike many of the more spectacular dated and ornamented pattern end houses in southwestern New Jersey, Mannington's few surviving eighteenth-century brick houses exhibit more restrained ornamentation. Most such ornamentation consists of a wall or walls laid in glazed-header Flemish bond or a projecting belt course of alternating plain and glazed headers, as seen on the Kiger house, or an inverted V-shaped line of glazed headers echoing the rake of the roof, as on the William Nicholson house.

Built in 1727 as a two-story dwelling with two end fireplaces, the John Pledger Jr. house typifies the "best" brick hall-parlor dwellings in southwestern New Jersey in terms of location and genealogy, ornamentation, original floor plan, and subsequent lateral growth. Located just outside of Salem facing Fenwick Creek, this dwelling occupies what was once a central location for transporting goods via water. Like so many other early brick houses, the Pledger house was home to one of the important founding families of Salem County. John Pledger Jr., the original owner of the house, was the son of one of the first English settlers in Fenwick's colony. In 1769 the house was sold to Robert Johnson II, a Pledger relative. Johnson's son, Robert Gibbon Johnson—the very same prominent landowner, local historian, and agricultural reformer who wrote the extensive treatise on marsh reclamation mentioned earlier—was born in this building.

The house consists of a gambrel-roofed two-story main block with a lower two-story gable-roofed kitchen wing attached to its western gable end (Figures 4.12 and 4.13). The original main block was a single-pile, three-bay, brick dwelling with two end chimneys and a gambrel roof.[73] This portion of the house is laid in glazed header Flemish bond on all four elevations and finished on all sides with a projecting stepped water table and checkered belt course. A brick cartouche suitable for a datestone adorns the west gable peak. Like most other hall-parlor dwellings, the Pledger house consisted of two unequal-sized rooms on the first floor, each of which was furnished with a gable fireplace and lit by two windows (Figure 4.14). Doors on both the front and rear elevations led into the larger of the two rooms, and a stair, which has since been reworked, led from this room to the second floor. When the present kitchen wing was joined to the west gable, the traditional open-plan arrangement of space was simply extended with an additional room on each floor. By the time Yorke photographed the house in 1887, both brick sections were visible, but the entire house was painted white, and the kitchen wing, which had two frame additions to its south elevation, had already lost its bake oven. Twentieth-century changes to the interior have been profound, even since the house was recorded by the Historic American Buildings Survey in the 1930s (Figure 4.15), and little original interior fabric remains. Today the house retains its original massing and its exterior features except for a few changes (Figure 4.16).[74]

While the Pledger house lacks a prominently dated, initialed, or patterned gable, it shares several other features common to elite brick houses in southwestern New Jersey and throughout the greater Delaware valley region. Typically, the most ornamented and formal end of a patterned brick dwelling was the parlor gable rather than the hall gable. When such houses received subsequent additions, which usually took the form of lower one- or one-and-one-half-story kitchen wings, these additions were typically joined to the hall side of the building, leaving the dates and the patterned brickwork of the parlor gable completely exposed (Figure 4.17). When additions were joined to a dated and initialed gable, they were built several feet lower than the main block—low enough so that the dated brickwork was still visible (Figure 4.18).

The Pledger house grew the same way many other elite brick houses in the region grew: its parlor gable was left exposed and its brick kitchen wing was joined to the opposite gable. A one-story kitchen addition may have been intended for this location from the outset. Although all four elevations of the main block are laid in glazed-header Flemish bond, the glazed headers on the exposed portion of the hall gable appear only at second-floor level, where they would have been seen above a projected one-story addition (Figure 4.19).[75] Built in 1722, the Abel Nicholson house in nearby Elsinboro Township reveals a similar shift in brickwork on the gable abutting the addition, and the evidence from several other early elite dwellings just across the river in Delaware likewise implies that future additions may have been planned for long before they were built.[76] The suggested placement for these intended additions not only retained time-honored spatial arrangements, by dictating the way these buildings should (and would) grow and change through time, but it also preserved the most symbolic and monumental aspects of these buildings. This constructed intentionality—the same intentionality that led the builders and owners of so many of these houses to date and ornament their parlor gables and to orient their houses so prominently—extended the monumental qualities of this elite South Jersey landscape well beyond the period of their construction. Another early image of the Pledger house suggests how pervasive this monumentalizing process was. When the artist J. A. Simkins painted the Pledger house in the late nineteenth century, he rendered a glazed header date on the western gable, but no architectural evidence for such an inlaid date exists (Figure 4.21). While the inclusion of such a detail in Simkins's painting may have been merely fanciful on the artist's part, it dramatically underscores just how important the monumental qualities of these buildings were.

Like the Pledger house, the William Hall Jr. and Richard Brick houses, built in 1724 and circa 1750, respectively, also represent some of the best brick hall-parlor plan houses in Mannington and share several common features. The houses are located in close proximity to each other near once-navigable tributaries of Mannington Creek. Both have primary facades laid in ornamental glazed header Flemish bond, and both consist of two-story hall-parlor main blocks with one or more lateral additions. And each dwelling represents one of Salem County's important founding families.

William Hall Jr. was the son of one of the wealthiest and most prominent landowners in Salem County. When his father died, around 1718, William Jr. inherited 1,000 acres in upper Mannington and a considerable amount of land in Salem. In 1724, he built his two-story brick house on the Mannington land.[77] Even as late as the end of the eighteenth century, the considerable Hall patrimony maintained its hold on the Mannington landscape, for Edward Hall, who occupied the ancestral home in 1798, was, like his predecessors, one of the wealthiest men in Mannington. The main block of the Halls' brick house was a hall-parlor plan similar in configuration to that of the Pledger house (Figure 4.22), but unlike the Pledger house, the original front elevation of the Hall house is the only wall ornamented with glazed-header Flemish bond brickwork and a projecting water table. In 1790, the house received a shorter, lateral, brick kitchen addition that clearly shows in the background of Yorke's 1888 photograph (Figure 4.1). Like most of the other brick dwellings, the Hall house was built with its most ornamental elevation facing the most public approach. When surrounding transportation routes changed and the primary entrance to the house was reoriented to what was once the back of the house, door and window openings were also shifted, and some have since been shifted several times. The Hall house lacks a brick datestone cartouche or any additional glazed-header ornamentation except on the exposed gable, where a chevron of glazed headers echoes the line of the roof rake (Figure 4.23). If the opposite gable had any ornamental brickwork when it was built, it has since been stuccoed over.

The Richard Brick house (Figures 4.24–4.26) represents a similar configuration, location, and genealogy. Richard Brick, a well-to-do tanner and another of Mannington's elite landowners, achieved local prominence when he served as a constable and political leader during the early years of the Revolution. His brother John, who had occupied the house earlier in the eighteenth century, was also a major southwestern New Jersey landowner and was instrumental in promoting the separation of Cumberland County from Salem County in 1748.[78] The Brick house began life circa 1750 as a traditional hall-parlor plan dwelling. Like the Hall house, only one of its elevations is ornamented with glazed-header Flemish bond brickwork and a projecting water table. Although the Brick house also lacks the ornamental belt courses visible on the earlier houses, it does display a recessed brick datestone cartouche near the peak of the exposed gable. The Brick house grew laterally as well. A lower one-and-one-half-story kitchen wing with a facade that was also laid in glazed-header Flemish bond was added to the hall gable shortly after the house was built. While a third, even lower, brick lean-to addition was joined to the exposed gable of the earlier addition some time in the nineteenth century and clearly shows in Yorke's 1887 image (and in a twentieth-century photograph by Sickler), it was apparently demolished or incorporated within the slightly larger and most recent kitchen addition which now occupies the same location. A frame ell was added in the mid-twentieth century, but to the rear or least public elevation. Thus, even over the course of several changes in the early nineteenth century, the traditional aspects of both of these hall-parlor dwellings were maintained. Furthermore, sometimes these hall-parlor buildings with appended kitchen wings were the result, not of constructed intentionality, but of fully realized spatial configurations from the outset. Many of these extant kitchen wings may simply have replaced earlier kitchens that were built at the same time as the main block. Although clear indications are not visible in the Pledger, Hall, or Brick houses, some early hall-parlor plan dwellings in southwestern New Jersey and across the river in Delaware show evidence for first-period kitchen wings attached laterally to one gable.[79]

Far less widespread than hall-parlor dwellings were two-room houses with their two rooms oriented front-to-back rather than side-by-side. Known commonly today as double-cell dwellings, these buildings are often viewed as urban adaptations of traditional hall-parlor plans, because they tend to be associated with the narrow lot requirements of urban settings rather than with more rural locations. Surviving rural pattern brickwork houses in this form tend to be found closer to urban centers.[80] Still, they resemble the other open plan dwellings in that one entered directly into the primary living space of the house rather than into an unheated stair passage. Most such buildings are two bays wide and contain either back-to-back corner fireplaces or a large cooking fireplace on the gable with a corner fireplace abutting it at the partition. While few double-cell dwellings survive in their original form in Mannington, such houses were occasionally built in the area. One such example exhibits an original core consisting of a two-bay frame double-cell plan with back-to-back corner fireplaces (Figure 4.27). Because this original core has since been extended by a later frame addition and the entire dwelling has been re-sided, the house presents a very different appearance today. It is now a more-or-less

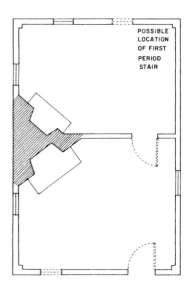

FIG. 4.27. Double-cell plans like this one consist of two rooms oriented front-to-back rather than side-by-side. They tend to be associated with urban settings, because of the narrowness of city lots.

symmetrically fenestrated five-bay double-pile dwelling with two end chimneys, largely concealing its original configuration.

While this frame dwelling and others like it hardly represent the level of monumentality achieved by the John Pledger and Richard Brick houses, the double-cell form was occasionally utilized in other, more prominent brick dwellings. Michael Chiarappa found several pattern brick double-cell houses in southwestern New Jersey. More significantly, because some builders favored a three-bay, window-door-window arrangement on the double cell form rather than the more commonly used two-bay scheme, Chiarappa has suggested that this three-bay opening arrangement, compressed as it sometimes was onto relatively narrow double-cell facades, had begun to carry significant status associations by mid-century, as it was increasingly linked throughout the region with the brick "plantation houses" of the native born elite.[81]

Other open plan types that were utilized in the region but are not well-represented in Mannington survivals included three- and four-room plans. The most common type of three-room plan is sometimes termed the Penn plan because of its description in William Penn's promotional literature. This "ordinary beginner's" house described in Penn's 1684 tract was a medium-sized house, 30 by 18 feet, with a partition near the middle and a second partition further dividing one end into two smaller rooms.[82] In terms of their interior spatial arrangement, three-room plan houses in west New Jersey resembled hall-parlor plans but with the parlor further subdivided into two rooms, often with back-to-back corner fireplaces. These houses averaged around 850–900 square feet on the first floor—larger than the typical hall-parlor house, which averaged around 420–680 square feet. Four-room plans were even less common. Such dwellings, which consisted of four rooms of unequal size on the first floor, appeared occasionally after the middle of the eighteenth century and were characterized by direct access into a heated room containing a stair. Such houses were typically more square in plan than the other open plan types and afforded more space on the first floor.[83]

Closed, or Georgian, forms that provided entry into an unheated stair passage began to appear in southwestern New Jersey in the mid-eighteenth century but were not widespread in the area. Local builders instead tended to favor a more traditional open plan. The distinguishing feature of Georgian form dwellings was the stair passage or entry, which provided an intermediate zone through which one gained access to other, more functionally specific rooms of the house. Such plan types created more clearly defined public and private zones and segmented once-open interior spaces. Georgian plans in the region were utilized in several arrangements including center-passage, side-passage, and half-passage plans. In a center-passage dwelling, the stair passage runs through the center of the house and provides access to two or four rooms on either side of the passage as well as to the second floor. In a side-passage dwelling, the stair passage runs along one gable and affords access to the room or rooms on the opposite side and to the second floor. Other, less commonly seen variations on the Georgian scheme included half-passage and corner-passage plans. A half-passage plan is characterized by a stair passage that penetrates only half the depth of the house, while the stair passage in a corner-passage plan dwelling has been enclosed and pushed into one corner. In Mannington, as in the broader southwestern New Jersey region, Georgian building conventions were gradually adapted, but usually only within the framework of the region's customary building practices.[84] Furthermore, many of the Georgian form buildings that did appear were based on accretion and were not the result of the original design.

A few buildings representing several of these Georgian plan types survive in Mannington. The original core of the William Smith house illustrates a typical two-thirds Georgian or side-passage plan (Figures 4.28 and 4.29). The core portion of the house was built circa 1760; it was extended on its passage elevation with a side-passage addition around 1850. The original core is now difficult to see from the exterior, as the structure is now a two-story, six-bay, rectangular frame dwelling with two end chimneys. Each floor of the original building consists of two squarish, evenly sized rooms next to a stair passage that runs the full depth of the house. The second floor is further subdivided with a small unheated room over the entry. The interior spaces feature a high level of finish, including fully paneled fireplace walls, chair rails, crown moldings, and a formal stair.

William Smith, the scion of yet another of Mannington's founding Quaker families, was a prominent landowner whose wealth and property placed him squarely in the ranks of the local elite. The original tract of Smith family land comprised 1,160 acres in 1685; although the landholding was much reduced by the end of the eighteenth century, the family patrimony remained significant. In 1797 William Smith was maintaining 180 acres of improved land; he owned seven horses and twenty cows, and his frame house in 1798 was valued at $1,000, nearly twice the value of most other dwellings and in the league of most brick houses in Mannington.

The core of the Smith house is a rare survival of a relatively intact eighteenth-century frame house and presents something of an odd case. Its interior finish displays craftsmanship of the highest order, and it is one of the most highly valued eighteenth-century frame houses in Mannington; yet, since Thomas Yorke photographed mostly first-period brick dwellings, it was not among the houses he monumentalized photographically in the 1880s. Likewise, Yorke's contemporary Thomas Shourds never mentioned the house in his genealogical accounts of Salem County's founding families, and Joseph Sickler's twentieth-century account also skipped over it. While this frame dwelling may have been just as economically and genealogically important as equivalently valued brick houses in its day, it was clearly elided from the region's ancestral map. In southwestern New Jersey, brick apparently carried additional cultural freight. It was the artificial, monumental, and durable qualities of brick that gave it a significant initial imprint on the landscape, but its significance was then seized upon by these late nineteenth-century chroniclers, who extended its importance well beyond the early eighteenth century. This elision of the

Smith house from the ancestral map extended even into the late twentieth century, as it was only recently "discovered" to be significant and to possess an early core.[85] Interestingly, the two surviving eighteenth-century frame houses included in this study were also transformed much more dramatically by later accretions than were most contemporary brick houses.

Some dwellings utilized plan configurations that display a more selective incorporation of the stair passage idea. The Hugh Middleton house, constructed in the mid-eighteenth century for a wealthy Quaker landowner, is a square two-story four-bay brick dwelling built in a half-passage configuration with the stair passage pushed to the rear of the house (Figures 4.30 and 4.31). As in the earlier open forms, one entered directly into the heated living space of the house rather than into an unheated passage, but an enclosed half-depth stair passage still channeled access to three of the four ground-floor rooms. The Middleton house is unusually deep and spacious compared to most other eighteenth-century dwellings in this area, but it, too, had received a lateral addition by the time Yorke photographed it. Its facade and one gable are laid in plain Flemish bond and ornamented with a stepped projecting belt course, while its rear elevation and opposite gable are laid in 3–5-course common bond. The interior of the house originally consisted of two roughly equal-sized rooms across the front and a half-depth center stair passage flanked by two smaller rooms. Back-to-back corner fireplaces set in an end-wall chimney heated the front and rear room on the northeast gable, and although the front room on the opposite gable had its own fireplace, the rear room behind it appears to have been unheated. A few similarly configured dwellings were built throughout southwestern New Jersey.[86] Still, closed Georgianized forms were far less common than the tradition-bound hall and hall-parlor arrangements which survive in such significant numbers today.

✦ ✦ ✦

The changes experienced by Mannington's brick architectural monuments thus honored the past and, in so doing, followed a common pattern. In the broader southwestern New Jersey region, as in Mannington, nineteenth-century changes made to many other early elite houses similarly respected the canon of the ancestral map by preserving customary spatial arrangements, underscoring genealogical linkages, and emphasizing initial symbolic inflections. Thomas Yorke's photographs capture the results of this ongoing dialogue with the past (Figures 4.32–4.36). While virtually all of the houses Yorke photographed in the 1880s had been expanded or altered in some way, most had retained their traditional spatial configurations, and most had had their most obvious ancestral linkages—glazed-brick dates and initials, datestone cartouches, and other ornamental brickwork—preserved during an extended process of change. When a roofline was raised from gambrel to gable or when a gable wing was added, the changes more often than not respected the symbolic features of the house and preserved the intentional prominence that was such a critical aspect of the initial building. This prominence or visibility—the monumental, landmark-like quality of these dwellings that was a function of their traditional design, their ornamental brickwork, their genealogical associations, and their high economic value, as well as their physical placement within the setting of a cultivated and improved tidewater landscape—carried symbolic freight from the day these houses were built. Thomas Shourds, writing in 1876, described how several different members of one founding family residing near Stow Creek had intentionally built their respective brick houses within the same visual compass, and he related one instance of "six brick dwellings, all in sight of each other, which were erected in the fore part of the last century."[87] Similarly, Alan Gowans has argued in his study of the mansions of Alloways Creek that, in South Jersey and "throughout the colonies, the idea of a founding family with an estate was a common ambition. And a 'great house'—great by comparison with its neighbors—was its commonest monument."[88]

By the end of the eighteenth century, these monumental family estates had acquired another quality that could only augment their genealogical and cultural importance: the patina of significant age. At a time when the new country was seeking to reinvent its identity, when a fresh array of symbols redolent of nationhood gained cultural importance, fine brick houses that represented important founding families and that were already nearly three-quarters of a century old offered tangible links to a local aristocracy and a New World "past." And by the time Shourds penned his regional history in 1876, this patina of age had attained new historical depth. Shourds betrayed something of the prevailing respect for the age of these buildings in his descriptions of "ancient" dwellings that had suffered untimely ends or that had had their walls carefully preserved within later remodelings.[89] Yorke's photographs document this same value system at work. But while this veneration for the past expressed by Shourds, Yorke, and their contemporaries was partly a product of Colonial Revival antiquarianism, it was also an extension and reappropriation of the initial landscape gesture inscribed by southwestern New Jersey's eighteenth-century elite. In short, it amounted to a late-nineteenth-century reformulation of the early-eighteenth-century ancestral map.

The significance of earlier landscape remnants to subsequent prevailing societies has been explored provocatively by Julia King in her archaeological work on the oligarchic society of nineteenth-century southern Maryland. King argues that, on the plantations that she studied, the ruins of certain earlier buildings were purposely maintained because they possessed symbolic value. These ruins "not only reminded the plantation owners and their families of the role played by their ancestors in the formation of the state and the nation, but they may also have subtly reinforced and justified a way of life that was failing." The remnants that were thus embedded in the landscape "may have served to justify a system that, through inheritance, kept land ownership out of reach for most people."[90] Simliarly, in her exploration of the linkages between early houses, house "myths," and American ideology, Anne Yentsch has argued that the structures built by eighteenth-century gentlemen and yeomen farmers are those that have survived through subsequent landscape reformulations and that it is these structures that are commonly perceived to be the homes of our ancestors. Yentsch maintains that both the omission from as well as the inclusion of specific family homes in local knowledge are intertwined acts in the formation of a local landscape ideology: "Family homes remembered and family homes forgotten are, in reality, two closely related elements in local historiography that tell of social structure and hierarchy within the community. They are equivalent acts; the logic behind them is the same. Superficially unrelated, remembering and forgetting factual details of home and land ownership accomplish the same ends."[91]

In the less-than-egalitarian society of eighteenth-century southwestern New Jersey, the dwellings built by the wealthy scions of founding families were intentionally exceptional and conspicuous. They both dominated their own landscape and set the stage for an ongoing dialogue with future generations. These monuments to lineage, property, and power were uncommon compared to the more modest dwellings inhabited by a predominantly landless population. But as these houses survived through the early years of nationhood into the first decades of the nineteenth century, they increasingly represented relics of an earlier era. These buildings, artifacts of mid-eighteenth-century power structures, had become aging housing stock in a changing landscape. Yet, like the vestigial pattern of large land parcels and the tide banks that burgeoned on the Mannington marshes, and like the tons of earth that were moved by enslaved African Americans on low country Georgia rice plantations, these dwellings endured as lasting imprints of a landed elite. The changes these houses experienced at the hands of succeeding generations remapped this earlier landscape. This redefinition of the ancestral landscape, like the "invented traditions" explored by Eric Hobsbawm and Terence Ranger, simultaneously historicized the past and attested to the continuing currency of a landscape expression that was mapped in the mid-eighteenth century.[92]

Enmeshed in these "ancient" buildings, then, an ongoing landscape ideology emerged that was at once a product of the present and an artifact of the past. And, like most ideologies, this particular landscape ideology was both pervasive and subtle. One of the key functions of ideology is to veil prevailing power relations by making them seem to be part of the natural order of things.[93] As in other maps, much of the power of the ancestral map inscribed by the South Jersey elite and reappropriated by subsequent generations inhered in its apparent truthfulness and neutrality. This power was expressed, not in terms of overt dominance, but as rhetoric. "Throttled down" to the material expressions of an eighteenth-century rural elite and redefined like the reoriented William Hall house, this ancestral map had acquired its own authority: it had become an invented tradition.

FIG. 4.5A. Lower Alloways Creek meetinghouse, Salem County, New Jersey, photographed in 1996. Many South Jersey Quaker meetinghouses resembled the earliest brick dwellings in the area.

FIG. 4.5B. Salem Friends meetinghouse, Salem, photographed in 1995. Like so many of the earlier dated brick dwellings in the area, this meetinghouse features a glazed-brick date on its gable wall.

FIG. 4.6. Kiger house or Jesuit Mission, Mannington, photographed in 1996.
Built early in the eighteenth century, the house overlooks Salem Creek,
a once-major water access route.

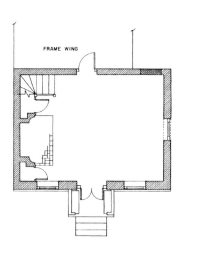
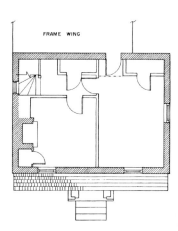

FIG. 4.7. Kiger house or Jesuit Mission, Mannington. Despite a typical size and spatial arrangement, this structure's ornamented brick construction distinguishes it from the norm.

FIG. 4.8. Kiger house or Jesuit Mission, Mannington Township. The single open room of the first floor was heated by a large fireplace. Historic American Buildings Survey; photographed May 3, 1939, by Nathaniel R. Ewan.

FIG. 4.9. William Nicholson house, Mannington, photographed in 1888. When the brick kitchen addition was built, the chevron pattern of glazed headers and the brick datestone cartouche on the gable wall of the original section were left visible. Photograph by Thomas Yorke; courtesy Salem County Historical Society.

FIG. 4.10. William Nicholson house, Mannington, photographed in 1996. The original front elevation, shown here, now faces cultivated farm fields rather than a landing on a navigable waterway. A second lateral addition, on the gable opposite the kitchen wing, further enlarged the dwelling.

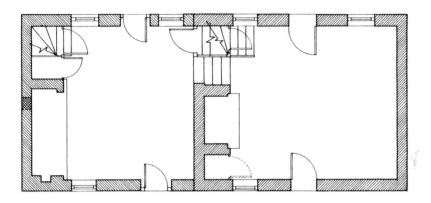

FIG. 4.11. William Nicholson house, Mannington, first-floor plan. The room on the right is from the first building period; the one on the left was added during the second period.

FIG. 4.12. John Pledger Jr. house, Mannington, photographed in 1888. Thomas Yorke's photographs of the Pledger house capture several additions and changes to the original building. Photograph by Thomas Yorke; courtesy Salem County Historical Society.

FIG. 4.13. John Pledger Jr. house, Mannington, photographed in 1888. By the 1880s, the Pledger house had lost the pent eave on its north (*facing*) elevation as well as its bake oven, evidence of which can be seen at the base of the west wall. Photograph by Thomas Yorke; courtesy Salem County Historical Society.

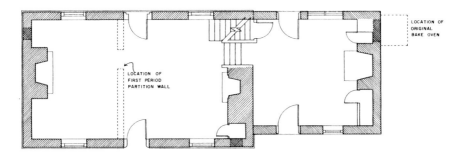

FIG. 4.14. John Pledger Jr. house, Mannington. Like most hall-parlor plan houses, this one had two unequal-sized rooms on the first floor.

FIG. 4.15. John Pledger Jr. house, Mannington. This mantel in the second-floor east bedroom #2 was recorded by the Historic American Buildings Survey. Photographed July 7, 1937, by Nathaniel R. Ewan.

FIG. 4.16. John Pledger Jr. house, Mannington, built in 1727, photographed in 1997. The view shows the south elevation and east gable end. The kitchen addition is on the opposite gable end. The ornamental Flemish bond brickwork was used only where it would be visible. Note how the brick bond shifts between the first and second floors where the original pent roof was attached.

FIG. 4.19. John Pledger Jr. house, Mannington, photographed in 1997. The ornamental glazed headers on the gable end of the main building were placed only where they would be seen above the kitchen addition. They are faintly visible here, beginnning six courses of brick below the bottom of the wooden pent eave.

FIG. 4.17. Chandler-Kent-Keasbey house, Quinton Township, Salem County, New Jersey, photographed in 1888. Note the original gambrel roofline on this dated gable. The Flemish bond brickwork from the first-period dwelling also remains visible at first-floor level near the corners of the facade. Photograph by Thomas Yorke; courtesy Salem County Historical Society.

FIG. 4.18. William Oakford house, Alloway Township, Salem County, New Jersey, photographed in 1888. A one-story addition once joined the dated gable, as evidenced by the whitish scar directly below the initials and date. The facade was ornamented with glazed-header Flemish bond brickwork. Photograph by Thomas Yorke; courtesy Salem County Historical Society.

FIG. 4.20. Abel Nicholson house, Elsinboro Township, Salem County, New Jersey. The Nicholson house was built in 1722, as attested by the date on the near gable end. Shifts in the brickwork on the far gable end of the original house suggest that a lower addition was planned from the start. The one shown here at far right was built in 1859. Historic American Buildings Survey; photographed September 21, 1936, by Nathaniel R. Ewan.

FIG. 4.21. J. H. Simkins, in his painting of the John Pledger Jr. house, embellished the building with an ornamental dated gable that apparently never existed. Courtesy Salem County Historical Society.

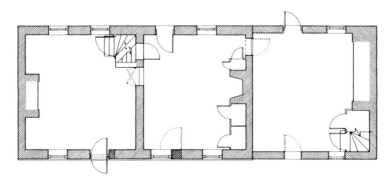

FIG. 4.22. William Hall house, Mannington, first-floor plan. The Hall house, owned by one of the most prominent men in Salem County, typifies the best hall-parlor brick dwellings in Mannington. The wall at the top of the drawing is now the front of the house (shown in Fig. 4.1); the opposite wall (see Fig. 4.23) was originally the front.

FIG. 4.23. William Hall house, Mannington, photographed in 1997. The most highly ornamented elevation of the Hall house (*at right*) is now the back of the house.

FIG. 4.24. Richard Brick (a.k.a. Dolbow) house, Mannington, northeast elevation. This house began life, ca. 1750, as a hall-parlor plan (*section at right*). A kitchen wing was added to the hall gable shortly after the house was built, and the lean-to addition at far left was built early in the nineteenth century. Historic American Buildings Survey; photographed, June 9, 1939, by Nathaniel R. Ewan.

FIG. 4.25. Richard Brick (a.k.a. Dolbow) house, Mannington, southwest elevation. Historic American Buildings Survey; photographed June 27, 1939, by Nathaniel R. Ewan.

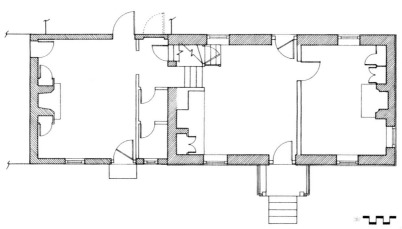

FIG. 4.26. Richard Brick (a.k.a. Dolbow) house, Mannington, first-floor plan of original main block and kitchen addition.

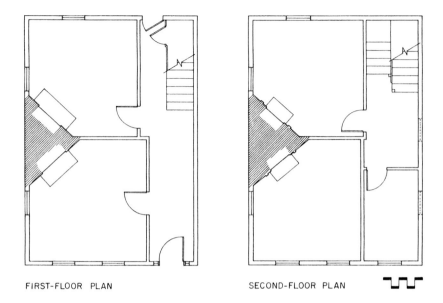

FIRST-FLOOR PLAN          SECOND-FLOOR PLAN

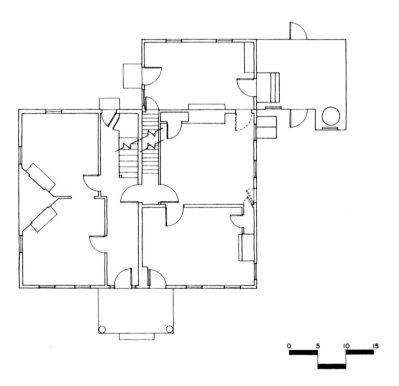

0    5    10    15

FIG. 4.28. William Smith house, Mannington. First-period first-floor (*top left*) and second-floor (*top right*) plans, and current first-floor plan (*bottom*).

FIG. 4.29. William Smith house, Mannington, photographed 1992. The left half of the house is the original section. Photograph by Bernard L. Herman.

FIG. 4.30. Hugh Middleton house, Mannington, photographed in 1888. Photograph by Thomas Yorke; courtesy Salem County Historical Society.

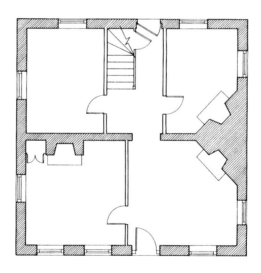

FIG. 4.31. Hugh Middleton house, Mannington, first-period first-floor plan, main block. The house originally consisted of four rooms arranged around a half-depth center stair passage.

FIG. 4.32. Richard Smith house, Elsinboro Township, Salem County, New Jersey, photographed in 1888. The original gambrel roof configuration and glazed initials, dates, and diamond patterns remain visible on the facing gable end. Photograph by Thomas Yorke; courtesy Salem County Historical Society.

FIG. 4.33. William Mecum house, Lower Penn's Neck Township, Salem County, New Jersey, photographed in 1888. When first built, the Mecum house had a gambrel roof. The original ornamental dates, initials, and roofline are visible on the gable wall at left. Photograph by Thomas Yorke; courtesy Salem County Historical Society.

FACING PAGE:

FIG. 4.34. Daniel Smith house, Quinton Battlefield, Salem County, New Jersey, photographed in 1888. Like many other early brick dwellings with ornamented gables, the Smith house was enlarged over time, but the changes respected the monumental quality of the original building by preserving the dated and patterned gables. Photograph by Thomas Yorke; courtesy Salem County Historical Society.

FIG. 4.35. Joseph Ware / Thomas Shourds house, Lower Alloways Creek Township, Salem County, New Jersey, photographed in 1888. The glazed date and original roof pitch of this eighteenth-century house remain visible above the later, frame addition. Photograph by Thomas Yorke; courtesy Salem County Historical Society.

FIG. 4.36. Samuel Tyler house, Salem, Salem County, New Jersey, photographed in 1888. When this house was enlarged, it also received a lateral addition with a slightly lower roofline. The ornamental brickwork and other architectural features of the original block were continued on the addition. Photograph by Thomas Yorke; courtesy Salem County Historical Society.

# CHAPTER 5

# A Region of Regions

A musician, William Priest, recalling his impressions upon visiting Philadelphia in the 1790s, remarked on the anticipatory building practices that were common in this city. "The first object of an industrious emigrant, who means to settle in Philadelphia," he wrote, "is to purchase a lot of ground in one of the vacant streets. He erects a small building forty or fifty feet from the line laid out for him by the city surveyor, and lives there till he can afford to build a house; when his former habitation serves him for a kitchen and wash-house. I have observed buildings in this state in the heart of the city; but they are more common in the outskirts." Referring to a particular individual who found himself in this very situation, Priest continued, "I am afraid it will be many years before he will be able to build in front."[1]

### SEAFORD, DELAWARE, 1830

With the following terse entry in his diary, William Morgan—doctor, farmer, Methodist preacher, keen observer of the denominational rivalry in his community, and charter member of the new Methodist Protestant Church in Seaford—documented the time-honored Sussex County practice of moving buildings: "That evening, all that was favorable to reform, i.e. all but the M.E. members met; a subscription was drawn up, and a sufficient amount subscribed, to justify the purchase of a meeting house about one and half mile out of town: and the next day the house was purchased, and in two weeks we sawed the house in two and placed four pair of wheels under one half with twelve yoake of oxen and five and twenty or thirty hands brought it safely into town. Set the two pieces ten feet apart and filled up the middle which made the house 38 by 24 feet."[2]

### SALEM COUNTY, NEW JERSEY, APRIL 1832

Robert Johnson of Salem itemized the expenses, labor, and materials involved in building a new brick washhouse and cistern on his property. In his accounting, Johnson noted that the building was constructed on a cellared stone foundation that measured 8½ feet deep. He recorded a total cost of $155.20, which included 35 days of labor to dig the cellar and cart dirt plus 6 days to dig the cistern and cart the bricks to build it. Additional charges covered both labor and materials: the mason's tenders, lime for the mortar, stone for the sink and window jambs, and stone jambs and an iron door for the ash hole. Before the new building could be begun, though, the old washhouse, which was standing on the same spot, needed to be pulled down.[3]

### LANCASTER COUNTY, PENNSYLVANIA, 1818

While English traveler and journalist William Cobbett marveled at the buildings and landscape improvements he encountered near Lancaster,

he found the way Pennsylvania farmers went about altering and improving their dwellings to be peculiar. "It is a curious thing to observe
the *farm-houses* in this country," he remarked. "They consist, almost without exception, of a considerably large and a very neat house, with sash
windows, and of a *small house,* which seems to have been *tacked on* to the
large one; and, the proportion they bear to each other, in point of dimensions, is, as nearly as possible, the proportion of size between a *Cow*
and *her Calf,* the latter a month old." "But," he wrote, "the process has
been the opposite of this instance of the works of nature, for, it is *the large
house which has grown out of the small one.*" Cobbett described how the
process of landscape improvement typically progressed. "The father or
grandfather, while he was toiling for his children, lived in the small
house, constructed chiefly by himself, and consisting of rude materials.
The means, accumulated in the small house, enabled the son to rear the
large one." While pride occasionally caused the younger generation to
demolish the older, smaller house, most retained the older buildings as
material evidence of their success. For, as Cobbett wrote, "few sons in
America have the folly or want of feeling to commit such acts of filial in-
gratitude, and of real self-abasement. For, what inheritance so valuable
and so honourable can a son enjoy as the proofs of his father's industry
and virtue?" Cobbett credited these domestic monuments, not just to
family success, but to America's political system itself. "The progress of
wealth and ease and enjoyment, evinced by this regular increase of the
size of the farmers' dwellings," he wrote, "is a spectacle, at once pleas-
ing, in a very high degree, in itself; and, in the same degree, it speaks
the praise of the system of government, under which it has taken place.
What a contrast with the farm-houses in England!"[4]

BETHLEHEM VICINITY, PENNSYLVANIA, 1825–1826

When Bernhard, Duke of Saxe-Weimar Eisenach, traveled to Bethlehem, Pennsylvania, in 1825–1826, he remarked on the recent changes
that seemed to be rippling through the area's predominantly German
community. The inhabitants, who were mostly descendants of German
emigrants, had retained their language, "although in an imperfect

state." They did this partly by printing German-language newspapers and almanacs for circulation among the German-speaking population, but also by not sending their children to school. Throughout the area, he noted, the inhabitants had betrayed these cultural values by investing heavily in their barns and churches but paying scant attention to their educational institutions. "In many villages where you see handsome brick buildings, stables, and barns," the duke wrote, "the school is a simple log-house, much worse than the school-houses I have seen among the Indians." Still, some change was beginning to occur, and, as the duke observed, this would eventually produce a marked effect on the built environment. "The difference is already perceptible in the state of Pennsylvania," he wrote. "They particularly complain that the former German farmers did not send their children to school at all; lately, however, they have become more ambitious, and attend the schools, because the legislature of Pennsylvania has passed a law, that no citizen shall sit on a jury unless he can read and write the English language. The German farmers consider it an honour to be called upon a jury, but find themselves deprived of that honour on account of their ignorance. They now, therefore, have their sons instructed in English."[5]

### PHILADELPHIA, PENNSYLVANIA, 1786

In the *Columbian Magazine* in 1786, Benjamin Rush described the "progress of population, agriculture, manners, and government" in Pennsylvania in a letter "to a friend in England." Settlement in Pennsylvania over time, Rush wrote, passed through three stages, each of which was characterized by a specific breed of farmer-settler moving from savagery to civilization and toward greater civic responsibility and participation. Moreover, each stage of settlement was visibly reflected in the built environment.

The object of the first settler was "to build a small cabbin of rough logs for himself and family. The floor of this cabbin is of earth," he continued, "the roof is of split logs—the light is received through the door, and, in some instances, thro' a small window made of greased paper." Because this first settler lived near Indians, he also acquired "a strong tincture of their manners." His behavior was marked by indulgence in violence and spirituous liquors; "his pleasures consist chiefly in fishing and hunting . . . and he eats, drinks, and sleeps in dirt and rags in his little cabbin." Because he abhorred laws and too much government, he repeatedly abandoned his settlement for the woods, where he began the process of clearing and building anew.

The improvements he left behind, however, were typically seized upon by the second species of settler, whose first object was "to build an addition to his cabbin," but, instead of rough logs, the second settler utilized "hewed logs: and as saw-mills generally follow settlements, his floors are made of boards; his roof is made of what are called clapboards." Moreover, since his house was larger, and contained more specialized spaces, the second-period settler reappropriated the original log cabin for use as a kitchen. This settler's next object was to clear meadow, plant an orchard, build a stable and barn, and increase his landholdings. Nonetheless, the house and farm of this species of farmer, while showing marked improvements over the first settler, "bear many marks of a weak tone of mind." His land failed to yield as much as it could, his animals performed "but half the labor that might be expected from them," and his windows stood unglazed, mostly, according to Rush, because he was rarely "a good member of civil or religious society," disinclined to contribute toward building a church or to support civil government. Typically, his accumulation of personal debts forced him to sell his plantation after a few years, often to a third breed of settler.

Usually "a man of property and good character," this third settler aimed to improve his soil as well as his farmscape by building a huge stone barn, "in some instances, 100 feet in front, and 40 in depth" but "made very compact" to keep his animals warm in winter and thereby consume less fuel. He used stoves to heat his house and further improved his holdings with well-kept fences, outbuildings, and a constant application of family labor. The final task of this settler was to build his own dwelling. This sometimes happened during his lifetime, but the task was "oftener bequeathed to his son, or the inheritor of his plantation: and hence we have a common saying among our best farmers, 'that a son should always begin where his father left off;' that is, he should begin his improvements, by building a commodious dwelling-house, suited to the improvements and value of the plantation." Rush described the house as usually stone, "large, convenient, and filled with useful and substantial furniture—It sometimes adjoins the house of the second settler, but is frequently placed at a little distance from it." Rush further declared that the third settler's material success bore a direct relationship to his role as a responsible member of civil society: "In proportion as he encreases in wealth, he values the protection of laws." This settler supported government, paid his taxes on time, and supported schools and churches because, as Rush observed, "benevolence and public spirit . . . are the natural offspring of affluence and independence." It was this class of conquering and virtuous farmers that constituted the majority of Pennsylvania's inhabitants and actually formed the backbone of its sociocultural identity. As Rush declared, "these are the men to whom Pennsylvania owes her ancient fame and consequence."[6]

In the waning years of the eighteenth century and the first few decades of the nineteenth, while the Delaware Valley remained an intensely local collection of distinct places, its built environment was beginning to experience perceptible change. As these contemporary accounts imply, some of these changes simply reflected the passage of time. In much of the region, already well into its second century of settlement, the landscape was beginning to evince cumulative layers of habitation. As existing building stock aged, grew ruinous, or outlived its usefulness in one location, it often needed to be replaced, reworked, or removed, like the old Salem County washhouse that Robert Johnson pulled down and replaced or the Delaware meetinghouse that William Morgan's congregation sawed in two, moved, and resurrected nearby. Some changes instead related to expectations for the future or shifting economic and geographic circumstances, such as the anticipatory and speculative building practices that, according to William Priest, shaped the development of much of Philadelphia's urban fabric. But, as several of these correspondents also suggest, many of these changes were linked to overlapping and far more complicated factors. Generational shifts, changing cultural and religious values, timing and length of settlement, ethnic intermingling, varying laws, economic and market forces, improved transportation networks, and the broader nationalizing transformations that swept through the early republic—all played a part in the landscape reediting they witnessed. For, as

William Cobbett remarked, it was not just the passage of time and generations that had so curiously transformed the "cow and calf" Lancaster County houses he described; it was also the prosperity and order encouraged by the American governmental system itself. Similarly, Benjamin Rush attributed Pennsylvania's attractive built environment as well as its very identity to the commitment to industry, law, and civic responsibility so characteristic of its virtuous "third species" of settler. And, as the Duke of Saxe-Weimar Eisenach observed, the cultural changes beginning to alter Bethlehem's German community, which also promised to transform its built environment, stemmed not so much from changes within the community itself but from the intersection of deeply held cultural values with new laws imposed from without. While such changes never effected the kind of sweeping and obvious transformations that would argue for considering these distinctly local environments as a cohesive "region," they did, gradually and subtly, begin to complicate the existing landscape by redefining existing categories of identity such as ethnicity, status, and belief.

The changes these observers commented upon may have ranged from subtle to dramatic, but they usually reflected choices, of individuals or communities or both. And, as many of these observers were quick to point out, each change to the built landscape also expressed, often overtly but sometimes almost imperceptibly, social and cultural values—values that were themselves often in a state of flux. In the minds of contemporary observers, such landscape changes were not only visible but often reciprocally linked to the way identity was expressed in the built environment. Although such expressions often occurred at the individual level, they were more frequently articulated in terms of shared community values, as in the Duke of Saxe-Weimar Eisenach's account of the changing educational values trickling through the German community near Bethlehem, or the choice of William Morgan's church congregation to purchase, move, and reerect a meetinghouse, or the "community" of speculative buildings that emerged on the expanding edges of Philadelphia in the early nineteenth century. And, although the communities that these correspondents referred to were clearly locally rather than regionally defined, to many, the national community was also beginning to play an increasingly important role in expressions of identity.

A regional identity, some scholars have argued, can emerge only after a national identity has been solidified. In other words, "regionalism" as a category is a relatively modern construct—one that is unstable, a by-product of nationhood, and often imposed from without rather than from within. And, because we are looking backward from our present-day vantage point within an established and enduring nation-state, we may tend to perceive a more stable, cohesive identity than might actually have existed in the past.[7]

The timing of the emergence of an American national identity has been a matter of considerable debate. While some scholars have argued that an American identity, inhering in the notion of American exceptionalism, extends back to the first encounters between Europeans and residents of the Americas at the end of the fifteenth century, others have maintained that the Civil War was actually the watershed event in the formation of a noticeable national identity and that prior to that crisis, national identity was weak or barely perceptible.[8] Michael Kammen suggests that the Civil War's stimulus to nationalism fostered a "growing reverence for familiar traditions that made loyalty meaningful and hence explicable."[9] And Wilbur Zelinsky has found that the Civil War constituted a key turning point in the maturation of the American nation-state and that visible landscape expressions of the central state's dominance over local, sectional, and agrarian interests and values emerged shortly thereafter. Since Americans did not have to contend with Europe's deeply layered cultural landscapes in their New World setting, they were able to approach their surroundings as a "blank canvas." The making of the New World landscape thus presented an exceptional situation for its inhabitants, but it was one that the American nation-state did not fully exploit until the nineteenth and twentieth centuries. Just as Zelinsky observes, and as we have seen demonstrated in the Delaware Valley, "local peculiarity was the rule in the landscapes of pre-Revolutionary European America, and remains so to a noticeable degree within such tracts to this day."[10]

"Local peculiarities" notwithstanding, the landscape underwent considerable change. The built environment was gradually altered, and the once "blank canvas" of the American landscape slowly began to acquire real historical depth. These landscapes often changed, as William Cobbett and the Duke of Saxe-Weimar Eisenach so astutely observed, in very particular ways. As the creolized dwellings and impressive bank barns of Lancaster County suggest so provocatively, it is the *way* these landscapes changed, and the *way* identities were gradually constructed and negotiated in both material and conceptual terms, from without as well as from within, that is most suggestive.[11]

In the localized landscapes of the Delaware Valley, as existing aspects of identity such as ethnicity, status, or race came to intersect or conflict with other forms of identity, they sometimes began to be redefined rather differently. In Lancaster County, while creolized Anglo-German dwellings may very well have represented a form of negotiation between "Anglo" and "German" ethnic categories, the newly important oversized bank barns, so closely linked with the Germanic population, may have also begun to eclipse older expressions of ethnicity, gradually assuming the symbolic place of land. In the conceptual border landscapes of early nineteenth-century North West Fork Hundred, fictional as well as factual accounts intertwined with the physical landscape to construct and amplify certain aspects of the region's identity, underscoring the centrality of boundaries and the consequences of transgressing those boundaries during an era when categories were hardening and distinctions between white and black, North and South, and slavery and freedom were becoming increasingly resonant. And, as New Jersey's "age of oligarchy" waned, the historicized landscapes of Salem County emerged as a provocative effort to preserve and map the physical remnants of an earlier, but fading, social order.

✦ ✦ ✦

While the Delaware Valley hardly exhibited a coherent regional identity in the early national period, a number of basic commonalities developed as the result of proximity and market orientation. Nearness to and regular exchange between the important urban markets of Philadelphia, Baltimore, and Lancaster constituted an important commonality among all of these landscapes. As the eighteenth century blended into the nineteenth, transportation networks improved, with the addition of new roads and, later, canals, and interregional intermingling increased. For example, as geographically isolated as North West Fork Hundred was, its extractive timber trade and that of the surrounding regions of Sussex County, Delaware, fed the more distant urban markets of Philadelphia and Baltimore as well as the nearby local markets just across the Delaware. References to this cross-river trade are common in newspapers and personal accounts, such as that of Richard Wood, a Greenwich, New Jersey, Quaker, who wrote in his diary one October day in 1810, "we have bought a load of shingles from Indian River [Delaware]."[12] But, because it encouraged the perpetuation of exploitative labor and land use practices, the demand imposed by these lucrative lumber markets in turn also permitted the continuation of the distinctive socioeconomic inequality that was so endemic to that particular Delaware landscape. Similarly, southwestern New Jersey farmers provided beef, dairy, and agricultural products to Delaware and Philadelphia markets across the river; and further to the west, Lancaster County farmers also maintained frequent contact with Philadelphia markets.[13] Richard Wood, of Greenwich, regularly shipped New Jersey produce, cheeses, and fatted cattle via ferry and schooner to markets in Delaware, New York, and Pennsylvania.[14] In the 1790s, the duc de la Rochefoucauld-Liancourt also remarked on the scope and interconnectedness of regional commerce, observing that in the countryside around Lancaster, there were "many mills, from which the flour is sent to Philadelphia in waggons. Returning, these waggons commonly bring merchandize, which is expedited from this place to every part of the back country."[15]

For Quakers throughout the Delaware Valley, the established system of monthly and yearly meetings promoted intraregional social and cultural as well as religious exchange. Richard Wood regularly traveled from his native Greenwich to monthly meetings in nearby Alloways Creek and to yearly meetings in Philadelphia. During the longer trips, he typically stopped along the way to visit friends, conduct business, or view the latest agricultural wonders, such as two merino rams "just arrived from Cadiz."[16]

Developing regional transportation arteries such as the Strasburg Road, built in 1784 to connect Philadelphia to Strasburg, Pennsylvania, and the Lancaster turnpike, which was begun in 1792 to connect Philadelphia and Lancaster, also increasingly linked people and places all along the way. Francis Baily described the regular stages that ran between Philadelphia and Baltimore in 1796 and 1797, and he marveled at the brand-new Lancaster turnpike, exclaiming, "There is, at present, but one turnpike-road on the continent, which is between Lancaster and Philadelphia, a distance of sixty-six miles, and is a masterpiece of its kind; it is paved with stone the whole way, and overlaid with gravel, so that it is never obstructed during the most severe seasons." And when Philip Fithian traveled by shallop from Philadelphia to his native Greenwich, New Jersey, the trip took thirty-one hours as compared to twelve hours by horseback and sixteen hours by stage, yet the ride across the Delaware between Port Penn and Cohansey Creek took only about half an hour via wherry. Although distant travel for most people remained time-consuming and sometimes arduous, it was a part of everyday life, and period accounts attest that it occurred frequently.[17]

Architecture, too, illustrates linkages that existed throughout the Delaware Valley and the surrounding region. For instance, the tradition of embellishing brick buildings with dated, initialed, or sometimes patterned gables, although it reached its most mature expression in southwestern New Jersey, appeared occasionally in Delaware and parts of Maryland. And as we have seen, because of proximity and because of shifting state boundary lines, numerous commonalities between building traditions in northwestern Sussex County, Delaware, and Maryland's Eastern Shore are widespread.[18] Similarly, the architectural intentionality that was practiced in southwestern New Jersey—the practice of subtly anticipating the location of lateral kitchen additions through the careful placement and/or elimination of ornamental glazed headers on the gable walls that would receive those additions—also existed in several eighteenth-century dwellings across the river in central Delaware. One-room plan dwellings like those enumerated in so many lower Delaware orphans court documents were nearly as common in southwestern New Jersey as they were in southern Delaware and on Maryland's Eastern Shore. The impressive two-story bank barns that contemporary observers enthused over were not totally confined to the Pennsylvania Piedmont, for architectural negotiations within the bank barn idiom also appeared as far south as north-central Delaware and well into Virginia.[19] And creolized German-Georgian house plans resembling those found in Warwick Township and the surrounding Lancaster County countryside were also built in Cumberland County and several other parts of Pennsylvania's German-settled backcountry.[20]

Although such linkages and commonalities clearly existed, the Delaware Valley landscapes examined here—along with numerous others throughout the area—retained a distinctly local character well into the early national period despite the kinds of changes that observers such as Cobbett, Rush, and Saxe-Weimar Eisenach commented upon. Some of this local distinctiveness began to blur as aging building stock was reworked, renegotiated, and replaced, as initial settlements matured, and as intraregional transportation and commerce increased; yet these landscapes remained local rather than regional, a mosaic of differing and often overlapping types of settlement cultures, architecture, and land use patterns rather than a monolith. Period observers even commented on this mosaic-like quality, for they repeatedly remarked upon the distinctions that the region's ethnic diversity created and the tendency for their contemporaries to stereotype cultural otherness. In 1791, Ferdinand-M. Bayard remarked that "the English, as well as the Americans, call indiscriminately the Germans and the Hollanders *the* Dutch." And as late as 1819, Henry Fearon echoed Saxe-Weimar Eisenach's comments when he remarked, "The original [German] language is still preserved, and there are even native Pennsylvanians who cannot speak the English language."[21] Some commentators went so far as to note the way this social and cultural diversity played itself out upon the landscape. In 1794, Theophile Cazenove actually mapped the distinct "layers" of Pennsylvania's population by drawing a series of semicircles to indicate distinct zones of settlement in the region. Cazenove's "first nucleus" consisted of Quakers, the "second layer" of Germans," the "third layer, beyond the Susquehanna, Irish and Scotch," and the "fourth layer, beyond the mountains, Irish, Scotch, New Englanders."[22] Such distinctions observed between populations and settlement cultures were mirrored by perceived differences in architectural and physiographic landscapes. Contemporary travelers regularly commented upon the dissimilarities of the various Delaware Valley landscapes they saw even as they passed through them, evincing acute awareness of distinctions between building traditions, farming practices, and soil types as well as the variety of manners and hospitality they encountered.

Conceptually, too, these were very different landscapes, for although these places were sometimes physiographically distinct from one another, they were also cultural creations that were both shaped by and perceived within distinct cultural categories. While the processes of assimilation and acculturation and the practice of historicizing the past undoubtedly contributed to the formation of these cultural landscapes, perceptions also played a significant role. Similarly, the less conscious process of solidifying a perceived regional identity, like that of the lower Delmarva Peninsula, through fictional and factual accounts played—and continues to play—a part. These cultural categories, like the ethnic landscapes encountered in Pennsylvania by Benjamin Rush and Johann Schoepf, the "marginal" landscapes of southern Delaware captured in and amplified by legend and literature, and the consciously ancestral landscapes of Salem County, New Jersey, that were recreated, perpetuated, and recanonized by subsequent generations, imparted meaning to past landscapes and continue to shape our interpretations of those landscapes today. In turn, this reflexivity or continuing exchange between past and present landscape perceptions and iterations constitutes a critical aspect of that slippery and deceptively monolithic-sounding concept known as a regional identity.

While the three subregions considered here broadly represent the diversity of the Delaware Valley cultural landscape, many other equally distinctive and localized environments contributed to the area's overall mosaic quality. Like the Pennsylvania German regions of Lancaster County and the Anglo-Quaker–dominated parts of southwestern New Jersey, the cultural landscapes of several of these areas have often been defined by the distinctiveness of their various settlement groups. For example, much of Willistown Township in Chester County, Pennsylvania, was located within the boundaries of the original Welsh Tract, a 40,000-acre tract of land laid out in 1682. The Welsh Tract was originally intended for Welsh settlers in Pennsylvania with the expectation that they would eventually become a separate barony, with the freedom to manage their own municipal affairs and retain their own language. Although this notion of a separate political entity proved impractical, the township was settled primarily by Welsh and English immigrants, and the area's many surviving eighteenth- and early-nineteenth-century buildings attest to the impact of their original imprint.

In some cases, material evidence of particular settlement groups survives in isolated pockets of specific building practices. For instance, in the Dutch-settled areas of the valley, including parts of Monmouth County, New Jersey, and Sussex County, Delaware, a distinctive type of timber framing known as H-bent or anchor-bent framing survives in a number of early buildings and confirms the lasting cultural imprint of early settlements.[23] Similarly, the New England–derived joinery traditions found in architecture from the earliest period of durable building in one pocket of Cumberland County, New Jersey, document the material impact of a significant early out-migration from Rhode Island and other parts of New England. A distinctive type of shouldered corner post that appears throughout parts of Rhode Island but only crops up in this one section of the Delaware Valley provides physical evidence of this important cultural and historical link.[24]

Several other parts of the Delaware Valley, because of their location, settlement patterns, or both, represent confluences of ethnic and architectural traditions rather than the kinds of homogeneous cultural landscapes represented by Warwick and Mannington Townships. For example, while much of Monmouth County, New Jersey, was affected by the Dutch imprint and looked primarily toward New York as its cultural and market center and toward the Atlantic Ocean or the Raritan River as its primary commercial arteries, the western part of the county occupied a midpoint between drainages as well as between two major developing urban centers. Located in the extreme southwestern part of the county, Upper Freehold Township and the immediately surrounding area, like North West Fork Hundred in Delaware, thus stood on both a geographic border and a market border. The township was closely tied with Delaware Valley transportation routes, linking it westward and southward, but because of its geographic location, settlement history, and proximity to more heavily Dutch-influenced areas, it was also historically and culturally linked to points north and east. The area's early settlement derived from the New England Baptist and Quaker communities originating in the older set-tled areas of Monmouth County, but it also drew some Scots, Dutch, French Huguenots, and English Quakers from Burlington and Philadelphia. The township's early heterogeneity as well as its geographic location at an inter-section between major commercial centers helped to establish the mixed identity of its built environment.[25]

Other areas in the Delaware Valley, such as the western and northern bor-ders of Chester County, Pennsylvania, also illustrate a heterogeneity that stands in marked contrast to more solidly English, Dutch, or German town-ships. Coventry Township, at the northern tip of Chester County where it bor-ders on the German-settled part of Berks County, exhibited a collision of cul-tural and architectural traditions. While some of its inhabitants were of British extraction, many others were descended from German, French, or Dutch ori-gins. This heterogeneity played out in the built environment, resulting in a mixture of building traditions and a series of complex negotiations among them.[26]

Like Coventry, Sadsbury Township, located on the westernmost border of Chester County and on the developing turnpike corridor between Philadel-phia and Lancaster, was already becoming heterogeneous by the end of the eighteenth century, but this heterogeneity was due to its geographic location as much as its settlement history. Populated by a mixture of Scots-Irish and En-glish Quaker settlers, the area still retained a strong English imprint, but it was located close enough to more strongly Germanic regions of Lancaster County area to be exposed to that area's culture and building practices through pre-vailing labor practices and everyday commercial exchange.

The distinctions among all of these various cultural landscapes mirror the social and cultural complexity of the entire Delaware Valley. The point that is worth underscoring here is that the local character of these places, individu-ally as well as collectively, argues against the notion of a monolithic regional commonality. The concept of a "Delaware Valley regional culture" such as the Quaker culture that some historians have postulated may be a useful contrivance for grasping certain cultural similarities or for comparing one part of the country to the whole, but such a monolithic construct has the net effect of simplifying and obfuscating the complexity of this landscape's constituent parts. The marginal cultural landscapes of southern Delaware were no more like those "Quaker" landscapes than they were like the Pennsylvania-German or Dutch-settled landscapes to the north and west. Even the historicized ancestral "Quaker" landscapes of southwestern New Jersey stood as localized expressions rather than broad regional commonalities. And although the "social flexibility" that some scholars have noted was one characteristic of greater Mid-Atlantic society, this broad similarity was offset by the equally distinctive differences that we have seen.

While the subregional landscapes of the Delaware Valley were increasingly experiencing rebuilding and change in the late 1700s and early 1800s, they remained locally distinct rather than regionally cohesive. Although clearly enmeshed in the broader regional economy and moving gradually toward a more regional and national orientation as populations burgeoned and cultural intermingling increased, these landscapes remained largely local in terms of their cultural expressions. From the English-settled Quaker landscapes of the flat New Jersey tidewater to the rolling fertile landscapes of the Pennsylvania German Piedmont, from the Chesapeake-influenced "country of slavery" in southern Delaware to the cultural and architectural crossroads at the northern and western fringes of Chester County, Pennsylvania, from the Welsh-settled landscapes just west of Philadelphia to the localized settlements of New Englanders in Cumberland County, New Jersey, to Philadelphia's rapidly expanding and intensely speculative northeastern edge, this was a landscape of many distinct places, a congeries of localized cultures rather than a coherent regional entity. In the early national period, the Delaware Valley was not a cohesive region but a region of regions, each with its own building traditions, settlement-based populations, land use patterns, material culture, and accompanying landscape perceptions. If "identity" and "region" imply unity, consistency, and commonality, then, at least in the material and documentary record, the regional identity of the Delaware Valley was defined, not by commonality, but by difference.

**TABLE 1**

*Mean Statistics for Warwick Township, 1798*

|  | Total Population (N=364) | German (92%) (N=335) | Non-German (8%) (N=29) |
|---|---|---|---|
| Total assessed property value | $2,391 | $2,473 | $1,444 |
| Total assessed acreage | 103 | 106 | 64 |
| % owner-occupied | 71% | 70% | 72% |
| Footprint square footage, primary dwelling | 711 (N=341) | 718 (N=315) | 636 (N=26) |
| Total square footage, all floors, primary dwelling | 1,028 (N=326) | 1,039 (N=301) | 891 (N=25) |
| Construction material, primary dwelling | 61% log | 60% log | 81% log |
|  | 28% stone | 29% stone | 11% stone |
|  | 2% frame and/ or plank | 2% frame and/ or plank | — |
|  | 5% stone and log | — | 4% stone and log |
|  | 1% brick | 1% brick | — |
|  | 3% unknown (N=345) | 2% unknown (N=319) | 4% unknown (N=26) |
| Footprint square footage, barn | 1,517 (N=243) | 1,517 (N=225) | 1,393 (N=16) |
| Construction material, barn | 45% log | 46% log | 41% log |
|  | 21% stone | 20% stone | 35% stone |
|  | 32% combination (stone and frame or stone and log) | 32% stone and log | 18% stone and log |
|  |  | 1% stone and frame | 6% stone and frame |
|  | 1% frame and plank | 0.5% frame and plank | — |
|  | 1% unknown (N=243) | 0.5% unknown (N=226) | — (N=17) |

SOURCE: 1798 Direct Tax, Schedules A and B

TABLE 2
*Mean Statistics for Hempfield Township, 1798*

| | Total Population (N=366) | German (79%) (N=287) | Non-German (21%) (N=75) |
|---|---|---|---|
| Total assessed property value | $2,339 | $2,557 | $1,518 |
| Total assessed acreage | 64 | 72 | 36 |
| % owner-occupied | 76% | 72% | 95% |
| Footprint square footage, primary dwelling | 735 (N=299) | 740 (N=241) | 710 (N=56) |
| Total square footage, all floors, primary dwelling | 1,012 (N=264) | 975 (N=215) | 1,156 (N=47) |
| Construction material, primary dwelling | 69% log 21% stone 3% combination 7% brick 2% frame or board 1% unknown (N=300) | 66% log 23% stone 3% combination 5% brick 1% frame and/ or board 2% unknown (N=242) | 66% log 14% stone 2% log and frame 13% brick 5% frame and/ or board (N=56) |
| Footprint square footage; barn | 1,796 (N=155) | 1,850 (N=135) | 1,431 (N=34) |
| Construction material, barn | 52% log 31% stone 15% combination 2% unknown (N=156) | 49% log 33% stone 16% combination 2% unknown (N=136) | 74% log 16% stone 11% combination (N=19) |

SOURCE: 1798 Direct Tax, Schedules A and B

TABLE 3
*Mean Statistics for Conestoga Township, 1798*

| | Total Population (N=146) | German (N=101) | Non-German (N=45) |
|---|---|---|---|
| Total assessed property value | $2,862 | $3,048 | $2,445 |
| Total assessed acreage | 100 | 103 | 94 |
| % owner-occupied | 89% | 60% | 62% |
| Footprint square footage, primary dwelling | 802 (N=136) | 823 (N=91) | 756 (N=43) |
| Total square footage, all floors, primary dwelling | 1,034 (N=134) | 1,091 (N=91) | 914 (N=43) |
| Construction material, primary dwelling | 69% log | 69% log | 70% log |
| | 24% stone | 23% stone | 26% stone |
| | 5% stone and log | 6% stone and log | 4% stone and log |
| | 1% brick | 1% brick | — |
| | 1% unknown | 1% unknown | — |
| | (N=137) | (N=94) | (N=43) |
| Footprint square footage, barn | 1,678 (N=111) | 1,649 (N=77) | 1,744 (N=34) |
| Construction material, barn | 26% log | 28% log | 21% log |
| | 13% stone | 17% stone | 2% stone |
| | 27% stone and log | 39% stone and log | 50% stone and log |
| | 15% stone below, log above | — | — |
| | 19% unknown | 17% unknown | 23% unknown |
| | (N=112) | (N=78) | (N=34) |

SOURCE: 1798 Direct Tax, Schedules A and B

TABLE 4

*Mean Statistics for Property Owners, Sadsbury Township, 1798*

| | Total Population (N=94) |
|---|---|
| Total assessed property value | $1,599 |
| Total assessed acreage | 129 |
| % owner-occupied | 83% |
| Footprint square footage, primary dwelling | 604 |
| Total square footage, all floors, primary dwelling | 1,050 |
| Construction material, primary dwelling | 39% log |
| | 58% stone |
| | 3% stone and log |
| Footprint square footage, first barn | 840 |
| Construction material, first barn | 78% log |
| | 4% frame |
| | 8% stone |
| | 5% stone and log |
| Square footage, springhouse | 143 |
| Construction material, springhouse | 9% log |
| | 88% stone |
| | 3% stone and log |

SOURCE: 1798 Direct Tax, Schedules A and B

**TABLE 5**
*Mean Statistics for Total Population, 1798 and 1815, Warwick Township*

| | 1798 (N=364) | 1815 (N=456) |
|---|---|---|
| Total assessed property value | $2,391 | $5,718 |
| Total assessed acreage | 103 | 76 |
| % tenant-occupied | 19% | data not available |
| Footprint square footage, primary dwelling | 711 (N=341) | 738 (N=410) |
| Total square footage, all floors, primary dwelling | 1,028 (N=326) | 1,055 (N=410) |
| Construction material, primary dwelling | 61% log | 56% timber |
| | 28% stone | 24% stone |
| | 5% stone and log | 4% combination (stone and wood, brick and wood, stone and timber, stone and brick) |
| | 2% frame and/or plank | 8% frame or wood |
| | 1% brick | 7% brick |
| | 3% unknown | 1% unknown |
| | (N=345) | (N=412) |
| Footprint square footage, barn | 1,517 (N=243) | 1,733 (N=250) |
| Construction material, barn | 45% log | 17% log, timber, or log and frame |
| | 21% stone | 29% stone |
| | 1% frame and plank | 18% wood |
| | 1% stone and frame | 24% first story stone, second story wood |
| | — | 1% first story stone, second story brick |
| | 31% stone and log | 8% combination (stone and wood, stone and timber, frame and boards) |
| | — | 1% brick |
| | 1% unknown | 2% unknown |
| | (N=226) | (N=17) |

SOURCE: 1798 Direct Tax, Schedules A and B; 1815 direct tax

TABLE 6
*Mean Statistics for Property Owners by Ethnic Origin, Warwick Township, 1815*

|  | German (81%) (N=368) | Non-German (19%) (N=88) |
|---|---|---|
| Total assessed property value | $6,330 | $3,157 |
| Total assessed acreage | 82 | 51 |
| Footprint square footage, primary dwelling | 766 | 610 |
| Total square footage, all floors, primary dwelling | 1,111 | 797 |
| Construction material, primary dwelling | 53% timber | 70% timber |
|  | 25% stone | 15% stone |
|  | 8% wood or frame | 8% wood or frame |
|  | 8% brick | 3% brick |
|  | 5% combination | 3% combination |
|  | 1% unknown | 1% unknown |
| Footprint square footage, first barn | 1,817 | 1,252 |
| Construction material, first barn | 14% log, timber, or log and frame | 37% log, timber, or frame and boards |
|  | 31% stone | 17% stone |
|  | 18% wood | 23% wood |
|  | 1% brick | — |
|  | 26% first story stone, second story brick | — |
|  | — | 15% first story stone, second story wood |
|  | 7% combination (stone and wood, stone and timber, frame and boards) | 5% combination |
|  | 3% unknown | 3% unknown |

SOURCE: 1815 Direct Tax

TABLE 7
*North West Fork Hundred: Distribution of Wealth, 1801, 1803, and 1816*

| Decile | % of Wealth Owned in 1801 | % of Wealth Owned in 1803 | % of Wealth Owned in 1816 |
|---|---|---|---|
| 1st | 2.79 | 1.49 | 1.18 |
| 2nd | 4.26 | 2.6 | 1.61 |
| 3rd | 4.78 | 3.3 | 1.92 |
| 4th | 5.86 | 3.7 | 2.14 |
| 5th | 6.54 | 4.27 | 2.51 |
| 6th | 7.6 | 5.1 | 3.28 |
| 7th | 8.02 | 7.13 | 5.2 |
| 8th | 10.75 | 10.53 | 9.75 |
| 9th | 15.68 | 18.04 | 17.86 |
| 10th | 33.72 | 43.84 | 54.55 |

SOURCE: 1801, 1803, and 1816 tax lists, North West Fork Hundred
NOTE: Total populations: 367 in 1801, 555 in 1803, 670 in 1816.

TABLE 8

*White, Slave, and Free African American Population in Sussex County, Delaware, 1800*

| Hundred | Total Population | % White | % Slave | % Free African American |
|---|---|---|---|---|
| Baltimore | 1,395 | 79.5 (1,107) | 18.5 (256) | 2 (32) |
| Broad Creek | 1,819 | 83 (1,511) | 13 (235) | 4 (73) |
| Broadkiln | 2,577 | 85 (2,182) | 10 (255) | 5 (140) |
| Cedar Creek | 2,515 | 78 (1,964) | 15 (382) | 7 (169) |
| Dagsborough | 1,418 | 75 (1,058) | 19 (270) | 6 (90) |
| Indian River | 1,547 | 73 (1,122) | 15 (240) | 12 (185) |
| Lewes & Rehoboth | 1,514 | 69 (1,045) | 16 (239) | 15 (230) |
| Little Creek | 2,184 | 84 (1,835) | 12 (255) | 4 (94) |
| Nanticoke | 1,872 | 81 (1,516) | 13 (239) | 6 (117) |
| North West Fork | 2,537 | 76.5 (1,940) | 18 (459) | 5.5 (138) |

SOURCE: Delaware Census, Manuscript Returns, 1800

TABLE 9

*White, Slave, and Free African American Population in Sussex County, Delaware, 1810*

| Hundred | Total Population | % White | % Slave | % Free African American |
|---|---|---|---|---|
| Baltimore, Broadkiln, Dagsborough, Indian River, and Lewes & Rehoboth | 10,107 | 76 (7,689) | 9 (948) | 15 (1,470) |
| Broad Creek | 3,789 | 85 (3,229) | 8 (289) | 7 (271) |
| Cedar Creek | 3,874 | 71 (2,768) | 8 (310) | 21 (796) |
| Little Creek | 3,840 | 83 (3,167) | 7 (281) | 10 (392) |
| Nanticoke | 2,843 | 80 (2,275) | 7 (192) | 13 (376) |
| North West Fork | 3,297 | 79 (2,619) | 12 (382) | 9 (296) |

SOURCE: Delaware Census, Manuscript Returns, 1810

TABLE 10

*White, Slave, and Free African American Population in Sussex County, Delaware, 1820*

| Hundred | Total Population | % White | % Slave | % Free African American |
|---|---|---|---|---|
| Baltimore | 2,057 | 83 (1,704) | 11 (228) | 6 (125) |
| Broad Creek | 2,599 | 80 (2,080) | 10 (254) | 10 (265) |
| Broadkiln | 2,731 | 84 (2,285) | 5 (144) | 11 (302) |
| Cedar Creek | 2,230 | 74 (1,648) | 5 (121) | 21 (461) |
| Dagsborough | 2,204 | 77 (1,694) | 8.5 (189) | 14.5 (321) |
| Indian River | 1,887 | 67 (1,260) | 9 (180) | 24 (447) |
| Lewes & Rehoboth | 1,657 | 75 (1,242) | 7 (111) | 18 (304) |
| Little Creek | 2,851 | 79 (2,247) | 14 (396) | 7 (208) |
| Nanticoke | 2,335 | 78 (1,828) | 8 (181) | 14 (326) |
| North West Fork | 3,408 | 75 (2,553) | 13 (440) | 12 (415) |

SOURCE: Delaware Census, Manuscript Returns, 1820

TABLE 11

*White, Slave, and Free African American Population in Sussex County, Delaware, 1830*

| Hundred | Total Population | % White | % Slave | % Free African American |
|---|---|---|---|---|
| Baltimore | 2,176 | 84.88 (1,847) | 9.46 (206) | 5.65 (123) |
| Broad Creek | 2,851 | 82.71 (2,359) | 8 (229) | 9.22 (263) |
| Broadkiln | 3,892 | 79.52 (3,095) | 2.87 (112) | 17.6 (685) |
| Cedar Creek | 2,737 | 72.89 (1,995) | 4.36 (119) | 22.84 (623) |
| Dagsborough | 2,551 | 73.53 (1,876) | 6.3 (161) | 20.14 (514) |
| Indian River | 1,936 | 71.28 (1,380) | 9.55 (185) | 19.16 (371) |
| Lewes & Rehoboth | 1,881 | 67.35 (1,267) | 6.69 (126) | 25.9 (488) |
| Little Creek | 3,207 | 79.79 (2,559) | 10.94 (351) | 9.26 (297) |
| Nanticoke | 2,366 | 78.14 (1,849) | 3.59 (85) | 18.25 (432) |
| North West Fork | 3,528 | 70.97 (2,504) | 9.75 (344) | 19.27 (680) |

SOURCE: Delaware Census, Manuscript Returns, 1830

TABLE 12

*Slave Ownership by Household in Sussex County, Delaware, 1800*

| Hundred | Total Households | Total Slaveowning Households | % Slaveowning Households | Total Free African American Households | % Free African American Households |
|---|---|---|---|---|---|
| Baltimore | 199 | 67 | 34 | 4 | 2 |
| Broad Creek | 340 | 89 | 26 | 15 | 4 |
| Broadkiln | 423 | 90 | 21 | 26 | 6 |
| Cedar Creek Hundred | 408 | 117 | 29 | 39 | 10 |
| Dagsborough | 221 | 68 | 31 | 19 | 8 |
| Indian River | 243 | 72 | 30 | 38 | 16 |
| Lewes & Rehoboth | 237 | 76 | 32 | 46 | 19 |
| Little Creek | 334 | 89 | 27 | 18 | 5 |
| Nanticoke | 303 | 68 | 22 | 22 | 7 |
| North West Fork | 356 | 119 | 33 | 25 | 7 |

SOURCE: Delaware Census, Manuscript Returns, 1800

TABLE 13

*Slave Ownership by Household in Sussex County, Delaware, 1810*

| Hundred | Total Households | Total Slaveowning Households | % Slaveowning Households | Total Free African American Households | % Free African American Households |
|---|---|---|---|---|---|
| Baltimore, Broadkiln, Dagsborough, Indian River, and Lewes & Rehoboth | 1,737 | 364 | 21 | 387 | 22 |
| Broad Creek | 503 | 82 | 16 | 79 | 16 |
| Cedar Creek | 534 | 120 | 22 | 166 | 31 |
| Little Creek | 487 | 78 | 16 | 90 | 18 |
| Nanticoke | 423 | 64 | 15 | 75 | 18 |
| North West Fork | 480 | 120 | 25 | 95 | 20 |

SOURCE: Delaware Census, Manuscript Returns, 1810

**TABLE 14**
*Slave Ownership by Household in Sussex County, Delaware, 1820*

| Hundred | Total Households | Total Slaveowning Households | % Slaveowning Households | Total Free African American Households | % Free African American Households |
|---|---|---|---|---|---|
| Baltimore | 305 | 55 | 18 | 24 | 8 |
| Broad Creek | 411 | 78 | 18.9 | 44 | 11 |
| Broadkiln | 460 | 65 | 14 | 102 | 22 |
| Cedar Creek | 398 | 63 | 15.8 | 121 | 30 |
| Dagsborough | 365 | 61 | 16.7 | 73 | 20 |
| Indian River | 312 | 58 | 18.5 | 107 | 34 |
| Lewes & Rehoboth | 264 | 51 | 19.3 | 80 | 30 |
| Little Creek | 436 | 114 | 26 | 51 | 12 |
| Nanticoke | 364 | 66 | 18.1 | 87 | 24 |
| North West Fork | 490 | 113 | 23 | 123 | 25 |

SOURCE: Delaware Census, Manuscript Returns, 1820

**TABLE 15**
*Slave Ownership by Household in Sussex County, Delaware, 1830*

| Hundred | Total Households | Total Slaveowning Households | % Slaveowning Households | Total Free African American Households | % Free African American Households |
|---|---|---|---|---|---|
| Baltimore | 346 | 50 | 14 | 31 | 9 |
| Broad Creek | 452 | 78 | 17 | 79 | 17 |
| Broadkiln | 661 | 66 | 10 | 189 | 29 |
| Cedar Creek | 470 | 55 | 12 | 174 | 37 |
| Dagsborough | 401 | 60 | 15 | 108 | 27 |
| Indian River | 300 | 48 | 16 | 80 | 27 |
| Lewes & Rehoboth | 323 | 59 | 18 | 107 | 33 |
| Little Creek | 513 | 111 | 22 | 88 | 17 |
| Nanticoke | 377 | 25 | 7 | 137 | 36 |
| North West Fork | 533 | 75 | 14 | 205 | 38 |

SOURCE: Delaware Census, Manuscript Returns, 1830

**TABLE 16**
*Landlessness in Mannington Township, 1773–1797*

| Date | Total Taxable Population | Landless Population | % Landless |
|------|------|------|------|
| 1773 | 207 | 108 | 52 |
| 1779 | 254 | 122 | 48 |
| 1795 | 263 | 118 | 45 |
| 1797 | 251 | 103 | 41 |

SOURCE: Mannington Local Tax Assessments for 1773, 1779, 1795, and 1797. A local tax list for 1789 also survives, but a significant portion of the population was apparently overlooked by the tax assessor, and the figures for numbers of landowners and landless individuals are so completely skewed when compared to the general trend exhibited in the other four lists that I have eliminated the 1789 list altogether from the comparison above. For example, according to the 1789 list, the total taxable population was 235, which is lower than the population in both 1779 and 1795. Also, this list enumerates the holdings of a total of 188 landed individuals, far more than were listed a decade earlier, and far more than were exhibited in both subsequent tax assessments. Similarly, the degree of landlessness shown by this list is much lower, both in terms of raw numbers (47) and in terms of percentages of the total population (20%), than on the previous and subsequent lists.

**TABLE 17**
*Distribution of Wealth in Mannington Township, 1797*

| Decile (N=25 per decile) | % Total Wealth (converted to pence) | Cumulative % | Number of Owners/Occupants of Brick Houses (N=22) |
|------|------|------|------|
| 1 | 1.4 (678) | 1.4 | 0 |
| 2 | 1.5 (759) | 2.9 | 0 |
| 3 | 2.7 (1,319) | 5.6 | 0 |
| 4 | 3.0 (1,500) | 8.6 | 0 |
| 5 | 3.6 (1,813) | 12.2 | 0 |
| 6 | 6.3 (3,109) | 18.5 | 0 |
| 7 | 10.4 (5,158) | 28.9 | 3 |
| 8 | 14.6 (7,262) | 43.5 | 5 |
| 9 | 21.5 (10,680) | 65.0 | 4 |
| 10 | 35.0 (17,433) | 100.0 | 10 |

SOURCE: Mannington Local Tax Assessment for 1797
N=250

TABLE 18

*Assessed Wealth Patterns Among Owners and Occupants of Brick Dwellings,*
*1797–1798*

| Portion of Population | Mean Value of Dwelling and Lot (1798 "A" list only; dwellings not listed or valued on 1797 list) | Mean Pound Value* | Mean Tax Assessed |
|---|---|---|---|
| Owners and occupants of brick houses only (N=22) (1797) (N=20) (1798) | $842.50 | £128 | £2..5..3 |
| Total landholding population only (N=148) | $536.45 | £72 | £1..4..8 |
| Total population including householders, single men, single men with horse, and negroes (N=251) | — | £42 | £0..16..1 |

SOURCE: Mannington 1797 Local Tax, cross-linked to Mannington 1798 Direct Tax, "A" List.
The 1797 Mannington list is calculated in pounds, shillings, and pence, while the 1798 direct
tax is in dollars and cents.
*"Pound value" is listed only on the 1797 local tax list and appears for propertied individuals
only. For "Householders, Single Men, Single Men with Horse, and Negroes," only names and
assessed taxes were listed.

TABLE 19

*Land and Livestock Ownership Patterns among Owners and*
*Occupants of Brick Dwellings, 1797–1798*

| Portion of Population | Mean Number of Improved Acres | Mean Value of Improved Acreage | Mean Number of Unimproved Acres | Mean Value of Unimproved Acreage | Mean Number of Horses | Mean Number of Cattle |
|---|---|---|---|---|---|---|
| Owners and occupants of brick houses only (N=22) | 193 | £69.09 | 15 | £2.95 | 4 or 5 | 13 or 14 |
| Total landholding population only (N=148) | 125 | £56.03 | 5 | £1.48 | 2 or 3 | 7 or 8 |
| Total population including householders, single men, single men with horse, and negroes (N=251) | 73 | £33.04 | 3 | £0.87 | 1 or 2 | 4 or 5 |

SOURCE: Mannington 1797 Local Tax, cross-linked to Mannington 1798 Direct Tax, "A" List

TABLE 20

*Percentages of Improved Land in Salem County, New Jersey, 1815–1823*

| Township or Borough | 1815 | 1818 | 1819 | 1823 |
|---|---|---|---|---|
| Salem | 100 | 100 | 100 | 100 |
| Elsinboro | 78 | 78 | 78 | 84 |
| Lower Alloways Creek | 68 | 68 | 68 | 68 |
| Upper Alloways Creek | 78 | 72 | 70 | 73 |
| Pittsgrove | 45 | 45 | 45 | 80 |
| Pilesgrove | 97 | 98 | 97 | 97 |
| Upper Penn's Neck | 91 | 98 | 98 | 98 |
| Lower Penn's Neck | 99 | 99 | 99 | 95 |
| Mannington | 97 | 97 | 97 | 98 |

SOURCE: Abstracts for Salem County, New Jersey, 1815, 1818, 1819, 1823

TABLE 21

*Assessed Wealth Patterns among Mannington Meadow Company Property Owners, 1797*

| Portion of Population | Mean Value | Mean Number of Improved Acres | Mean Value of Improved Acres | Mean Number of Unimproved Acres | Mean Value of Unimproved Acres | Mean Tax Assessed |
|---|---|---|---|---|---|---|
| Mannington Meadow Company property owners 1797–1799 (N=14) | £151 | 204 | £76.43 | 11 | £3.21 | £2..12..7 |
| Total landholding population only (N=148, 69%) | £72 | 125 | £56.03 | 5 | £1.48 | £1..4..8 |
| Total population including householders, single men, single men with horse, and negroes* (N=251) | £42 | 73 | £33.04 | 3 | £0.87 | £0..16..1 |

SOURCES: Mannington Meadow Company Minute Book and Mannington 1797 Local Tax.

TABLE 22

*Assessed Wealth Patterns among Owners and Occupants of*
*Brick Dwellings in Lower Alloways Creek, 1798*

| Portion of Population | Mean Total Value | Mean Number of Acres | Mean Square Footage of Dwelling (all floors) | % of Population with a Barn | Mean Square Footage of Barn | % of Population with Kitchen Separately Enumerated | % of Population Consisting of Owner-Occupants |
|---|---|---|---|---|---|---|---|
| Owners and occupants of brick houses only (N=22, 7% of pop.) | $2,619.55 | 144 | 1,154 | 82 | 812 | 95 | 73 |
| Total landholding population (N=313) | $731.48 | 68 | 654 | 24 | 682 | 26 | 61 |

SOURCE: Lower Alloways Creek 1798 Direct Tax, "A" and "B" List (cross-linked)

TABLE 23

*Assessed Wealth Patterns among Owners and Occupants of*
*Brick Dwellings in Pittsgrove, 1798*

| Portion of Population | Mean Value of Dwelling and Lot* | Mean Value† | Mean County Tax Assessed, 1798 Local List |
|---|---|---|---|
| Owners and occupants of brick houses only (N=12) | $588 | £40 | £0..14..0 |
| Total landholding population only (N=268) | $279 | £18 | £0..7..0 |
| Total population including householders, single men, single men with horse, and negroes (N=386) | — | £12 | £0..5..0 |

SOURCE: Pittsgrove 1798 Local Tax, cross-linked to Pittsgrove 1798 direct Tax, "A" List. The 1798 local tax list is calculated in pounds, shillings, and pence, while the 1798 direct tax is in dollars and cents.
* From 1798 "A" list only; dwellings not listed or valued on 1798 local list.
†Value is listed only on the 1798 local tax list and appears for propertied individuals only. For "Householders, Single Men, Single Men with Horse, and Negroes," only names and assessed taxes were listed.

TABLE 24
*Land and Livestock Ownership Patterns among Owners and*
*Occupants of Brick Dwellings in Pittsgrove, 1798*

| Portion of Population | Mean Number of Improved Acres | Mean Value of Improved Acreage | Mean Number of Unimproved Acres | Mean Value of Unimproved Acreage | Mean Number of Horses | Mean Number of Cattle |
|---|---|---|---|---|---|---|
| Owners and occupants of brick houses only (N=12) (3% of total pop.; 4% of landholding pop.; 7% of pop. listed on 1798 "A" list) | 158 | £23.91 | 15.83 | £1.33 | 2 or 3 | 5 or 6 |
| Total landholding population only (N=268) | 73 | £14 | 64* | £1.59* | 1 or 2 | 3 or 4 |
| Total population including householders, single men, single men with horse, and negroes (N=386) | 51 | £9.76 | 44* | £1.10* | 1 or 2 | 2 or 3 |

SOURCE: Pittsgrove 1798 Local Tax, cross-linked to Pittsgrove 1798 Direct Tax, "A" List

* Four individuals who do not appear at all on the "A" list for the 1798 Direct Tax held very large parcels (over 600 acres each) of unimproved acreage in the township. These individuals were not assessed for any livestock. When these outliers are removed and the averages recalculated, the mean number of unimproved acres drops to 50 for members of the landholding population, with a mean value of £1.58 per acre, and 35 for the total population, with a mean value of £1.09 per acre.

# NOTES

## ⊹ ABBREVIATIONS USED IN NOTES AND SOURCES ⊹

| | |
|---|---|
| CCA | Chester County Archives |
| CCHS | Chester County Historical Society |
| DSA | Delaware State Archives |
| HSP | Historical Society of Pennsylvania |
| LCHS | Lancaster County Historical Society |
| SCCH | Salem County Court House |
| SCHS | Salem County Historical Society |
| SCOC | Sussex County Orphans Court |

## ⊹ INTRODUCTION ⊹

1. Dell Upton, "The City as Material Culture" in Anne Elizabeth Yentsch and Mary C. Beaudry, eds., *The Art and Mystery of Historical Archaeology: Essays in Honor of James Deetz* (Boca Raton, Fla.: CRC Press, 1992), 51–53.

2. Peirce Lewis, "Common Landscapes as Historic Documents," in Steven Lubar and W. David Kingery, eds., *History from Things: Essays on Material Culture* (Washington, D.C.: Smithsonian Institution Press, 1993), 116.

3. William Least Heat-Moon, *Prairy Earth (a deep map)*. (Boston: Houghton Mifflin, 1991), 14–15.

4. James Lemon's 1972 study of early southeastern Pennsylvania and David Hackett Fischer's more sweeping work on the persistence of British folkways in America illustrate two of the most recent attempts to provide a broad and comprehensive view of the Delaware Valley as a single region. James T. Lemon, *The Best Poor Man's Country: A Geographical Study of Early Southeastern Pennsylvania* (New York: W. W. Norton, 1976). David Hackett Fischer, *Albion's Seed: Four British Folkways in America* (New York: Oxford University Press, 1989).

5. Wilbur Zelinsky, "North America's Vernacular Regions," in Wilbur Zelinsky, *Exploring the Beloved Country: Geographic Forays into American Society and Culture* (Iowa City: University of Iowa Press, 1994), 410–436, quote on 424. Zelinsky defines a vernacular region as "neither something created by government, corporate, or journalistic fiat,

nor the scientist's artifact, however sophisticated or otherwise, contrived to serve some scholarly or pedagogic purpose." He quotes Terry Jordan in explaining that such regions "'are those perceived to exist by their inhabitants and other members of the population at large. They exist as part of the popular or folk culture. Rather than being the intellectual creation of the professional geographer, the vernacular region is the product of the spatial perception of average people. Rather than being based on carefully chosen, quantifiable criteria, such regions are composites of the mental maps of the population.'" Zelinsky, 410, quote on 434, n. 2. Zelinsky is quoting directly from Jordan, "Perceptual Regions in Texas," *Geographical Review* 68 (1978): 293–307, quote on 293.

6. The Pennsylvania German community in the early national period was itself an extremely diverse one, representing fractured religious affiliations and a wide assortment of local backgrounds. Nonetheless, contemporaries often used the term "German" to describe this population, even though Pennsylvania Germans were an extraordinarily diverse group. For the purpose of expediency throughout this volume, the term "German" is often used in place of the more cumbersome but technically correct "Pennsylvania German" despite the tremendous diversity of the group.

7. Fred B. Kniffen, "Folk Housing: Key to Diffusion," in Dell Upton and John Michael Vlach, eds., *Common Places: Readings in American Vernacular Architecture* (Athens: University of Georgia Press, 1986), 3–26.

8. Michael O'Brien, "On Observing the Quicksand," *American Historical Review,* October 1999, www.historycooperative.org/journals/ahr/104.4/aho01202.html, p. 6 (accessed June 22, 2002). Others have made this same point about the emergence of regional identity in Europe. See Celia Applegate, "A Europe of Regions: Reflections on the Historiography of Sub-National Places in Modern Times," *American Historical Review,* October 1999, www.historycooperative.org/journals/ahr/104.4/aho01157.html, pp. 4, 10, 30 (June 22, 2002). Benedict R. Anderson, *Imagined Communities: Reflections on the Origins and Spread of Nationalism* (London: Verso, 1983, 1996).

9. Edward L. Ayers and Peter S. Onuf, "Introduction," in Edward L. Ayers, Patricia Nelson Limerick, Stephen Nissenbaum, and Peter S. Onuf, *All Over the Map: Rethinking American Regions* (Baltimore: Johns Hopkins University Press, 1996), 1–10.

10. Patricia Limerick has argued that, "for all the differences in the associations they bring to mind, both region and section [as categories of historical analysis] have been running an equal risk of obsolescence in the late twentieth century. In an era of historical inquiry in which the categories of race, ethnicity, class, and gender have taken center stage, both region and section register as relics and antiques." Patricia Nelson Limerick, "Region and Reason," in Edward L. Ayers, Patricia Nelson Limerick, Stephen Nissenbaum, and Peter S. Onuf, *All Over the Map: Rethinking American Regions* (Baltimore: Johns Hopkins University Press, 1996), 83–104, quote on 84. O'Brien, "On Observing the Quicksand," 4.

CHAPTER 1

"The motley middle" is Michael Zuckerman's phrase. Michael Zuckerman, "Introduction: Puritans, Cavaliers, and the Motley Middle," in Zuckerman, ed., *Friends and Neighbors: Group Life in America's First Plural Society* (Philadelphia: Temple University Press, 1982), 3–25. I am indebted to Patricia Keller for her many helpful comments on an earlier version of this chapter.

1. This opening sequence is inspired by James Deetz's work. James Deetz, *In Small Things Forgotten: An Archaeology of Early American Life* (New York: Anchor/Doubleday, 1996), 1–4. 1815 Direct Tax, Warwick Township, Lancaster County, Pa., LCHS. "Northern Liberty Improvements. Building Lots for Sale." March 11, 1803, In Robert Brooke, Survey Notes, AM 26711, #122, 2 vols., HSP. Samuel Wilson Account Books, vol. 2:66, CCHS. Gabrielle Milan Lanier, "Samuel Wilson's Working World: Builders and Building in Chester County, Pennsylvania, 1780–1827" (Master's thesis, Winterthur Program in Early American Culture, University of Delaware, 1989), 20–23. Manlove Adams Probate Inventory, February 10, 1806, DSA. Sussex County Orphans Court Extracts: Manlove Adams L-204 (1813) and L-275 (1814). Mannington Meadows Company Minute Book, 1756 to 1828, MS #70, SCHS.

2. Thomas Cooper, *Some Information Respecting America* (London: printed for J. Johnson, 1794), 16–17, 58–59. Isaac Weld, *Travels Through the States of North America* (London: William L. Clements, 1800), 116.

3. James T. Lemon, *The Best Poor Man's Country: A Geographical Study of Early Southeastern Pennsylvania* (New York: W. W. Norton, 1976), 49–51.

4. As twenty-first-century observers looking backward, we enjoy a vantage point that is both privileged and artificial. While hindsight may make some past regional distinctions seem clear to us, to the historical actors and inhabitants of these locales, such distinctions might not have seemed so precise. Their distinctions, too, might not have been the same as ours. Regional boundaries have a way of shifting with time and circumstance, and the emergence of regions as self-conscious cultural entities is intimately linked to other factors. As Stephen Nissenbaum has noted, "a region, much like a class, is something that gets generated in the process of distinguishing itself from something else." Stephen Nissenbaum, "New England as Region and Nation," in Edward L. Ayers, Patricia Nelson Limerick, Stephen Nissenbaum, and Peter S. Onuf, *All Over the Map: Rethinking American Regions* (Baltimore: Johns Hopkins University Press, 1996), 38–61, 46. David M. Wrobel, "Beyond the Frontier-Region Dichotomy," *Pacific Historical Review* 65 (1996): 401–429, 409.

5. Michael Steiner and Clarence Mondale, *Region and Regionalism in the United States: A Source Book for the Humanities and Social Sciences* (New York: Garland Publishing, 1988), x. John Fraser Hart, "Presidential Address: The Highest Form of the Geographer's Art," *Annals of the Association of American Geographers* 72, no. 1 (March 1982), 1–29. For a focus on regional and landscape self-consciousness outside of North America, see also the work of Marc Bloch and W. G. Hoskins. Marc Bloch, *French Rural History: An Essay on Its Basic Characteristics,* translated by Janet Sondheimer (Berkeley: University of California Press, 1966. W. G. Hoskins, *The Making of the English Landscape* (New York:

Viking Penguin, 1955). Edward L. Ayers, Patricia Nelson Limerick, Stephen Nissenbaum, and Peter S. Onuf, *All Over the Map: Rethinking American Regions* (Baltimore: Johns Hopkins University Press, 1996), 2.

6. Hart, "Highest Form of Geographer's Art," 9–10. Robert L. Dorman, *Revolt of the Provinces: The Regionalist Movement in America, 1920–1945* (Chapel Hill: University of North Carolina Press, 1993), xiii.

7. Raymond Williams, *The Country and the City* (New York: Oxford University Press, 1973), 45. Michael C. Steiner, "Regionalism in the Great Depression," *Geographical Review* 73, no. 4 (October 1983), 437, 432–433.

8. Dorman, *Revolt of the Provinces*, xii–xiii, 34–35.

9. Howard W. Odum and Harry Estill Moore, *American Regionalism: A Cultural-Historical Approach to National Integration* (New York: Henry Holt, 1938), 2.

10. Henry Glassie, *Pattern in the Material Folk Culture of the Eastern United States* (Philadelphia: University of Pennsylvania Press, 1968), 34.

11. Glassie, *Pattern in Material Folk Culture,* 39.

12. Hans Kurath, "Dialect Areas, Settlement Areas, and Culture Areas in the United States," in Caroline F. Ware, ed., *The Cultural Approach to History.* New York: Columbia University Press, 1940), 331–345.

13. Edward L. Ayers and Peter S. Onuf, "Introduction," in Edward L. Ayers, Patricia Nelson Limerick, Stephen Nissenbaum, and Peter S. Onuf, *All Over the Map: Rethinking American Regions* (Baltimore: Johns Hopkins University Press, 1996), 3.

14. Fred B. Kniffen, "Folk Housing: Key to Diffusion," in Dell Upton and John Michael Vlach, eds., *Common Places: Readings in American Vernacular Architecture* (Athens: University of Georgia Press, 1986), 3–26.

15. Glassie, *Pattern in Material Folk Culture,* 36. Wilbur Zelinksy, "The Changing Character of North American Culture Areas," in Glen E. Lich, ed., *Regional Studies: The Interplay of Land and People* (College Station: Texas A & M University Press, 1992), 113–135, 113.

16. Robert Blair St. George, "Artifacts of Regional Consciousness in the Connecticut River Valley, 1700–1780," in Robert Blair St. George, ed., *Material Life in America, 1600–1800* (Boston: Northeastern University Press, 1988), 336. Here St. George is drawing from the work of Odum and Moore, *American Regionalism,* 29–34.

17. For example, Gerald Pocius has shown that daily sharing of resources and space is one of the defining factors for the region he studied in Calvert, Newfoundland. Gerald L. Pocius, *A Place to Belong: Community Order and Everyday Space in Calvert, Newfoundland* (Athens: University of Georgia Press, and Montreal: McGill-Queen's University Press, 1991). Ary Lamme argues that power often defines a landscape and that the control of landscapes is based as much on powerful ideas as it is on powerful people. Ary J. Lamme III. *America's Historic Landscapes: Community Power and the Preservation of Four National Historic Sites* (Knoxville: University of Tennessee Press, 1989). Michael Chiarappa maintains that the English Quaker dominance of the southwestern New Jersey landscape was related to well-established kinship networks. Michael J. Chiarappa, "'The First and Best Sort': Quakerism, Brick Artisanry, and the Vernacular Aesthetics

of Eighteenth-Century West New Jersey Pattern Brickwork Architecture" (Ph.D. diss., University of Pennsylvania, 1992).

18. Wilbur Zelinsky, *The Cultural Geography of the United States* (Englewood Cliffs, N.J.: Prentice-Hall, 1973), 76. James Lemon has also emphasized the difficulty of defining early American regional boundaries with any precision. See James T. Lemon, "Spatial Order: Households in Local Communities and Regions," in Jack P. Greene and J. R. Pole, eds., *Colonial British America: Essays in the New History of the Early Modern Era* (Baltimore: Johns Hopkins University Press, 1984), 86–122, 88–89.

19. Anthony P. Cohen, *The Symbolic Construction of Community* (London: Routledge, 1989), 102, 44, 46. Edward A. Chappell, "Acculturation in the Shenandoah Valley: Rhenish Houses of the Massanutten Settlement" in Dell Upton and John Michael Vlach, eds., *Common Places: Readings in American Vernacular Architecture* (Athens: University of Georgia Press, 1986), 27–57.

20. Cohen, *Symbolic Construction of Community*, 12. See also D. W. Meinig, ed., *The Interpretation of Ordinary Landscapes: Geographical Essays* (New York: Oxford University Press, 1979), 3, 34; and Yi-Fu Tuan, *Topophilia: A Study of Environmental Perception, Attitudes, and Values* (New York: Columbia University Press, 1974).

21. Zelinksy, "Changing Character of Culture Areas," 113.

22. Anthony P. Cohen, "Culture as Identity: An Anthropologist's View," *New Literary History* 24, no. 1 (Winter 1993): 195–209.

23. While an extended examination of the intersections of gender and regional and national identity is beyond the scope of this study, a number of recent historical and geographical studies explore aspects of this topic. See for example Mary Beth Norton, *Founding Mothers and Fathers: Gendered Power and the Forming of American Society* (New York: Alfred A. Knopf, 1996); Linda K. Kerber, *Women of the Republic: Intellect and Ideology in Revolutionary America* (Chapel Hill: University of North Carolina Press, 1980); Gillian Rose, *Feminism and Geography: The Limits of Geographical Knowledge* (Minneapolis: University of Minnesota Press, 1993); Mona Domosh and Joni Seager, *Putting Women in Place: Feminist Geographers Make Sense of the World* (New York: Guilford Press, 2001); and Linda McDowell, *Gender, Identity, and Place: Understanding Feminist Geographies* (Minneapolis: University of Minnesota Press, 1999). Sally McMurry's examination of one particular backcountry Pennsylvania landscape in *From Sugar Camps to Star Barns: Rural Life and Landscape in a Western Pennsylvania Community* (University Park: Pennsylvania State University Press, 2001) considers many of the intersections between gender, labor, and the landscape. Steven M. Nolt's recent book, *Foreigners in Their Own Land: Pennsylvania Germans in the Early Republic* (University Park: Pennsylvania State University Press, 2002) focuses on the Pennsylvania German encounter with American political and religious culture and argues that Pennsylvania Germans were the first major group to experience what he calls "ethnicization-as-Americanization."

24. Edward Ayers and Peter Onuf argue that regions themselves have historically been shifting constructions controlled by multiple factors but that a regional identity tends to be defined by unity and that such an identity is "usually more about belonging than it is about exclusion." Ayers and Onuf, "Introduction," 4, 9–10. The phrase

"imagined communities" is from Benedict R. Anderson, *Imagined Communities: Reflections on the Origins and Spread of Nationalism* (London: Verso, 1983, 1996).

25. Douglas Greenberg, "The Middle Colonies in Recent American Historiography, " *William and Mary Quarterly*, 3rd ser., 36 (1979): 396–427, 408–409. For excellent summaries of Middle Colonies historiography, see both Greenberg's 1979 article and the more recent follow-up by Wayne Bodle, "Themes and Directions in Middle Colonies Historiography, 1980–1994," *William and Mary Quarterly*, 3rd ser., 51, no. 3 (July 1994): 355–388.

26. Martyn J. Bowden, "Culture and Place: English Sub-Cultural Regions in New England in the Seventeenth Century," *Connecticut History* 35 (1994): 68–146, 68. The term "invented tradition" comes from Eric Hobsbawm and Terence Ranger, eds., *The Invention of Tradition* (Cambridge: Cambridge University Press, 1983). An invented tradition, as defined by Hobsbawm and Ranger, is "a set of practices, normally governed by overtly or tacitly accepted rules and of a ritual or symbolic nature, which seek to inculcate certain values and norms of behaviour by repetition, which automatically implies continuity with the past."

27. For examples of such redrawn regional distinctions, see Patricia U. Bonomi, "The Middle Colonies: Embryo of the New Political Order," in Alden T. Vaughan and George Athan Billias, eds., *Perspectives on Early American History: Essays in Honor of Richard B. Morris* (New York: Harper and Row, 1973), 63–94; Greenberg, "The Middle Colonies"; Bodle, "Themes and Directions"; and Zuckerman, "Puritans, Cavaliers, and the Motley Middle." See also Wayne Bodle, "The 'Myth of the Middle Colonies' Reconsidered: The Process of Regionalization in Early America," *Pennsylvania Magazine of History and Biography* 113, no. 4 (October 1989), 525–548. In this article, Bodle summarizes an earlier argument, developed in Robert Gough, "The Myth of the 'Middle Colonies'": An Analysis of Regionalization in Early America," *Pennsylvania Magazine of History and Biography* 107 (1983), 393–419. Historian Gough has argued for a middle *section* which consisted of two distinct "regions," the first of which would have included "New York, parts of western Connecticut, eastern New Jersey, and the northeast corner of Pennsylvania." The other would have contained "most of Pennsylvania, part of Maryland, and all of western New Jersey and Delaware." Gough also maintained that although "these 'regions' were adjacent they did not embrace or even face each other" but instead "looked outward, to the north and south, from their point of conjunction in central New Jersey." Wayne Bodle countered that, while the "mythical" middle region might, in fact, have existed, it probably makes more sense to view regions as "locuses of interactive behavior" rather than "continuous bundles of characteristics— whether identical, substantially similar, or merely comparable." Bodle's point is that "however different or diverse its parts may have been, if a constellation of identifiable places 'acted' like a region, it just may have been one." Bodle, "'Myth of the Middle Colonies' Reconsidered," 527–528. Gough, "The Myth of the 'Middle Colonies,'" 394–395; 404–406.

28. Zuckerman suggested that, in ethnic diversity, in politics, in economic action and attitude, in social mores, and in religion, "the Middle Atlantic exhibited more fully than any other colonial region the shape of things to come. New England and the

South presented fragments, at best, of the nineteenth-century synthesis. The middle colonies presented the synthesis itself." Zuckerman, "Puritans, Cavaliers, and the Motley Middle," 7, 11–12, 13, 15.

29. Greene maintained that "in their predominant pattern of settlement 'on isolated farmsteads and in crossroads communities,' in their housing and in their farm buildings, in their agricultural practices, in their diet, and in their willingness to move, they [the Middle Colonies] all displayed an astonishing similarity." Their tolerance for diversity and receptivity to change "provided the basis for an emerging cohesiveness and sense of identity that . . . promoted habits of 'social flexibility'" and permitted them to pass through the eighteenth century in relative "peace and harmony amidst diversity." Jack P. Greene, *Pursuits of Happiness: The Social Development of Early Modern British Colonies and the Formation of American Culture* (Chapel Hill: University of North Carolina Press, 1988), 5, 47–50, 52–53, 124–135, 137–138, 141.

30. Fisher argued that settlers who migrated from East Anglia to Massachusetts, from the south and west of England to Virginia, from the North Midlands to the Delaware Valley, and from the border regions between England, Scotland, and Ulster to the American borderlands left indelible imprints upon the folkways of each region that have lasted to the present. According to Fischer, influential immigrant elites consciously attempted to create colonial regional cultures resembling the regional cultures from which they had migrated. David Hackett Fischer, *Albion's Seed: Four British Folkways in America* (New York: Oxford University Press, 1989), 7, 419–603. See also David Hackett Fischer, "*Albion* and the Critics: Further Evidence and Reflection," *William and Mary Quarterly*, 3rd ser., 48, no. 2 (April 1991): 260–308.

31. Fischer, *Albion's Seed*, 424.

32. Fisher's study is an important one, not so much for what it does best, which is to argue for the significance of cultural continuities and synthesize much of the existing scholarship into an impressive narrative whole, but also for the research agendas its controversial argument has suggested and for the heated scholarly dialogue it has provoked. For cogent critiques of *Albion's Seed* in general and his treatment of the Delaware Valley in particular, see Jack P. Greene, "Transplanting Moments: Inheritance in the Formation of Early American Culture," *William and Mary Quarterly*, 3rd ser., 48, no. 2 (April 1991): 224–230; and Barry Levy, "Quakers, the Delaware Valley, and North Midlands Emigration to America," *William and Mary Quarterly*, 3rd ser., 48, no. 2 (April 1991): 246–252. For additional critiques of *Albion's Seed* and Fischer's response to them, see the "Forum" in *William and Mary Quarterly*, 3rd ser., 48, no. 2 (April 1991): 224–308. See also Robert D. Mitchell, "Review of *Albion's Seed: Four British Folkways in America*," *Virginia Magazine of History and Biography* 99 no. 1 (January 1991), 96.

33. To Fischer, who attributed all of the area's building practices to the Quaker influence, the architecture of the Delaware Valley was noticeably different from buildings in more northerly places, such as Massachusetts and Connecticut, and represented "a distinct regional vernacular." He characterized the region's houses as primarily built of stone and representative of "a plain style that emerged from Quaker-Pietist values and North Midland traditions." But while stone buildings did become more widespread in a few parts of southeastern Pennsylvania by the fourth quarter of the eigh-

teenth century, not until very late in the century did stone actually eclipse wood as the favored building material in these areas. For example, in Sadsbury Township, in Chester County, Pennsylvania, stone did not overtake log as the dominant building material for dwellings until 1798–1799. The overwhelming majority of dwellings throughout most of the Delaware Valley, including southeastern Pennsylvania and southwestern New Jersey, were built of log or frame from the earliest settlements and into the early nineteenth century. In fact, in southwestern New Jersey, stone dwellings were a decided rarity, and in lower Delaware, they were almost nonexistent. The 1798 federal direct tax, which was assessed well after the shift to the more durable materials that Fischer describes, indicates that stone houses constituted only .5 percent of all dwellings in Lower Alloways Creek, in Salem County, New Jersey, and only 12 percent of all dwellings were built of brick. Similarly, on the fertile German-settled Lancaster Plain, just 21–27 percent of all dwellings were built of stone. Only in certain regions of Chester County, Pennsylvania, did stone buildings predominate, representing roughly 52–58 percent of the total recorded dwelling stock. Fischer, *Albion's Seed,* 477, 476. Lanier, "Samuel Wilson's Working World," 34–36. 1798 Federal Direct Tax, Lower Alloways Creek, Salem County, N.J.; Hempfield, Warwick, and Conestoga Townships, Lancaster County, Pa.; Sadsbury and Tredyffrin Townships, Chester County, Pa.

Similarly, *Albion's Seed* associated certain building motifs and plan types with cultural continuities between the North Midlands of England and the Delaware Valley, but evidence for the regional specificity of these features in England as well as their clear linkage to New World offshoots remains far from conclusive. Pent roofs, which are shallow roofs projecting from walls and framed on extended joists between floors or at the base of a gable, are widespread in the Delaware Valley. *Albion's Seed* implied that pent roofs and door hoods constituted a significant material link between the parent culture and its New World offspring. Although pents and door hoods were certainly present on some North Midlands buildings and may, in fact, be related somehow to some sort of cultural preference, a clear link to the North Midlands region has never been firmly established. In addition, pents and door hoods were utilized in several other parts of England, including Wiltshire, Somerset, and Devonshire far to the south, and may also have constituted an urban amenity in market-related settings. Fischer, *Albion's Seed,* 478. Nat Alcock, personal communication, April 3, 1998. Bernard L. Herman, personal communication, March 27, 1998. For an example of an early-eighteenth-century dwelling with a door hood in Devon, well south of the North Midlands, see Eric Mercer, *English Vernacular Houses* (London: Her Majesty's Stationery Office, 1975), plate 39. For examples in Lancashire, see Royal Commission on the Historic Monuments of England, *Rural Houses of the Lancashire Pennines, 1560–1760* (London: Her Majesty's Stationery Office, 1985), plates 60 and 70. Pent roofs also appear on prints showing buildings in urban settings, as in William Hogarth's "Times of Day" series.

Finally, Fischer associated two plan types that appeared widely in the Delaware Valley, the three-room "Quaker Plan," which was first published in 1684 in William Penn's promotional literature, and the "four over four," with the North Midlands of England. Cary Carson et al., "Impermanent Architecture in the Southern American Colonies," *Winterthur Portfolio* 16, no. 2/3 (Summer–Autumn 1981), 141–144. Although the three-room plan has not been particularly well-documented in Britain and the overall distribution of the form as well as its frequency of use is not yet clearly understood, examples of this type do show up in regions far from the North Midlands, such as Suffolk. Although Fischer conceded that "other elements were later added" to these distinctly Quaker building ways, including Germanic *fachwerk* construction and house plans, Swiss bank houses, and a distinctive urban architectural style that developed in Philadelphia, he nonetheless emphasized that the dominant Quaker values "cultivated the plain style in sturdy structures that were designed for use rather than display" and exerted the most profound influence on an emerging regional style. Fischer, *Albion's Seed,* 481. Bernard L. Herman, personal communication, March 27, 1998.

34. Barry Levy, *Quakers and the American Family: British Settlement in the Delaware Valley* (New York: Oxford University Press, 1988), 6, 11.

35. Stephanie Grauman Wolf, *Urban Village: Population, Community, and Family Structure in Germantown, Pennsylvania, 1683–1800* (Princeton: Princeton University Press, 1976), 7.

36. Sally Schwartz, *"A Mixed Multitude": The Struggle for Toleration in Colonial Pennsylvania.* (New York: New York University Press, 1987), 3, 7.

37. Lemon, *Best Poor Man's Country,* xiii–xvi, 15–23. See also James T. Lemon, "The Agricultural Practices of National Groups in Eighteenth-Century Southeastern Pennsylvania," *Geographical Review* 56, no. 4 (October 1966): 467–496.

38. Peter O. Wacker, *Land and People: A Cultural Geography of Preindustrial New Jersey: Origins and Settlement Patterns* (New Brunswick, N.J.: Rutgers University Press, 1975), 408–412.

39. Peter O. Wacker and Paul G. E. Clemens, *Land Use in Early New Jersey: A Historical Geography* (Newark: New Jersey Historical Society, 1995), 35, 47, 263n. See also Peter O. Wacker, "Patterns and Problems in the Historical Geography of the Afro-American Population of New Jersey, 1726–1860," in Ralph E. Ehrenberg, ed., *Pattern and Process: Research in Historical Geography* (Washington, D.C.: Howard University Press, 1975), 25–72; and Peter O. Wacker, "The Dutch Culture Area in the Northeast, 1609–1800," *New Jersey History* 104, no. 1/2 (Spring–Summer 1986), 1–21.

40. Zelinsky maintained that the midland culture area, which first assumed its unique form in southeastern Pennsylvania, was marked by a prosperous agricultural society followed quickly by a mixed economy involving both mercantile and industrial activity. By the middle of the eighteenth century, much of the region had become significantly urban in character and was far ahead of contiguous areas to the north and south. Zelinsky, *Cultural Geography,* 125–126.

41. Zelinsky, *Cultural Geography,* 126–128.

42. Henry Chandlee Forman's early studies of Maryland and Virginia architecture exemplify this tradition of emphasizing documentation and description. See for example, Henry Chandlee Forman, *Old Buildings, Gardens, and Furniture in Tidewater Maryland* (Cambridge, Md.: Tidewater Publishers, 1967); Henry Chandlee Forman, *The Virginia Eastern Shore and Its British Origins: History, Gardens, and Antiquities* (Easton, Md.: Eastern Shore Publishers' Associates, 1975); and Henry Chandlee Forman, *Tidewater Maryland Architecture and Gardens* (New York: Architectural Book Publishing,

1956). For Delaware Valley examples, see, for instance, Joseph Sickler, *The Old Houses of Salem County* (Salem, N.J.: Sunbeam Press, 1949); Eleanor Raymond, *Early Domestic Architecture of Pennsylvania* (Exton, Pa.: Schiffer Publishing, 1971); and Margaret Berwind Schiffer, *Survey of Chester County, Pennsylvania Architecture: 17th, 18th and 19th Centuries* (Exton, Pa.: Schiffer Publishing, 1976).

43. G. Edwin Brumbaugh, "Colonial Architecture of the Pennsylvania Germans," *Pennsylvania German Society Proceedings* 41 (Norristown, Pa., 1933), 6. Scott T. Swank, "The Architectural Landscape," in Scott T. Swank et al., *Arts of the Pennsylvania Germans* (New York: W. W. Norton, 1983), 21–22.

44. Robert C. Bucher, "The Continental Log House," *Pennsylvania Folklife* 12, no. 4 (Summer 1962): 14–19; Arthur J. Lawton, "The Pre-Metric Foot and Its Use in Pennsylvania German Architecture," *Pennsylvania Folklife* 19, no. 1 (Autumn 1969): 37–45; Arthur J. Lawton, "The Ground Rules of Folk Architecture," *Pennsylvania Folklife* 23, no. 1 (Autumn 1973): 13–19; Henry Glassie, "A Central Chimney Continental Log House," *Pennsylvania Folklife* 18, no. 2 (Winter 1968–69): 33–39. See also Robert C. Bucher, "Grain in the Attic," *Pennsylvania Folklife* 13, no. 2 (Winter 1962–3): 7–15.

45. Swank, "The Architectural Landscape," 34. See also Scott T. Swank, "The Germanic Fragment," "Proxemic Patterns," and "From Dumb Dutch to Folk Heroes" in Swank et al., *Arts of the Pennsylvania Germans*, 3–19, 35–60, 61–76.

46. Philip E. Pendleton, *Oley Valley Heritage, The Colonial Years: 1700–1775* (Birdsboro, Pa.: Pennsylvania German Society and the Oley Valley Heritage Association, 1994).

47. Charles Bergengren, "From Lovers to Murderers: The Etiquette of Entry and the Social Implications of House Form," *Winterthur Portfolio* 29, no. 1 (1994), 43–72. See also Charles Lang Bergengren, "The Cycle of Transformations in the Houses of Schaefferstown, Pennsylvania" (Ph.D. diss., University of Pennsylvania, 1988).

48. Jack Michel, "'In a Manner and Fashion Suitable to Their Degree': A Preliminary Investigation of the Material Culture of Early Rural Pennsylvania," in *Working Papers from the Regional Economic History Research Center,* ed. Glenn Porter and William H. Mulligan Jr., 5, no. 1 (1981): 1–83.

49. Bernard L. Herman, *Architecture and Rural Life in Central Delaware, 1700–1900* (Knoxville: University of Tennessee Press, 1987). Bernard L. Herman, *The Stolen House* (Charlottesville: University Press of Virginia, 1992).

50. J. Ritchie Garrison, Bernard L. Herman, and Barbara McLean Ward, eds. *After Ratification: Material Life in Delaware, 1789–1820* (Newark, Del.: Museum Studies Program, 1988).

51. Peter O. Wacker, "Traditional House and Barn Types in New Jersey: Keys to Acculturation, Past Cultureographic Regions, and Settlement History," in H. J. Walker and W. G. Haag, eds., *Geoscience and Man*, Volume 5: *Man and Cultural Heritage* (Baton Rouge, La.: School of Geoscience, Louisiana State University, 1974), 163–176.

52. Paul Love, "Pattern Brickwork in Southern New Jersey," *Proceedings of the New Jersey Historical Society* 73 (July 1955): 182–208. Alan Gowans, "The Mansions of Alloways Creek," in Dell Upton and John Michael Vlach, eds., *Common Places: Readings in American Vernacular Architecture* (Athens: University of Georgia Press, 1986), 367–393.

53. Henry Glassie, "Eighteenth-Century Cultural Process in Delaware Valley Folk Building," in Dell Upton and John Michael Vlach, eds., *Common Places: Readings in American Vernacular Architecture* (Athens: University of Georgia Press, 1986), 394–425. Glassie, *Pattern in Material Folk Culture*, 36–64.

54. In a popular book on contemporary American regions, reporter Joel Garreau placed most of the Delaware Valley within a region he called "the Foundry." The Foundry, he said, is "the linchpin of North American development," a region that is largely defined by what man has done to the environment as he has attempted to get ahead, rather than by the environment itself. On Garreau's maps of his nine North American regions, the Foundry is bounded by "Dixie" to the south and "The Bread-basket," "Quebec," and "New England" to the west and north. By this measure, southern Delaware falls right on the borderline between the Foundry and Dixie; Maryland's lower Eastern Shore, "with its chicken farms and insulated, isolated rural poverty, is clearly old-line Dixie." Joel Garreau, *The Nine Nations of North America* (Boston: Houghton Mifflin, 1981), 62, 66, 67, map following 204. When Wilbur Zelinsky mapped North America as fourteen vernacular regions, he included the bulk of the lower Delaware Valley in the core of his "Eastern" region. Wilbur Zelinsky, "North America's Vernacular Regions," in *Exploring the Beloved Country: Geographic Forays into American Society and Culture* (Iowa City: University of Iowa Press, 1994), 410–436. Fred Kniffen's map of diffusion routes and Henry Glassie's map of cultural regions indicate similar core areas. Glassie, *Pattern in Material Culture*, 38–39. Kniffen, "Folk Housing," 13, 21.

55. Zelinsky, "Changing Character of Culture Areas," 113, 123. Zelinsky, "North America's Vernacular Regions," 424.

56. François Alexandre Frédéric, duc de La Rochefoucauld-Liancourt, *Travels through the United States of North America, the country of the Iroquois, and upper Canada, in the years 1795, 1796, 1797; with an authentic account of Lower Canada,* 2 vols. (London: R. Phillips, 1799), 1:23.

57. Rayner Wickersham Kelsey, ed., *Cazenove Journal 1794: A Record of the Journal of Theophile Cazenove through New Jersey and Pennsylvania* (Haverford: Pennsylvania History Press, 1922), 69–70.

58. Weld, *Travels Through the States,* 3–4.

59. Thomas Anburey, *Travels Through the Interior Parts of America: In a series of letters. By an officer.* (London: William Lane, 1789), 279–280.

⁕ CHAPTER 2 ⁕

This chapter is based on an earlier essay of the same title published in *The Domestic Landscape in the Early Republic: Architecture and Material Culture in Eighteenth-Century America,* Bernard L. Herman and Michael Steinitz, eds. (Knoxville: University of Tennessee Press, forthcoming), and on "A Region of Regions: Local and Regional Culture in the Delaware Valley, 1780–1830" (Ph.D. diss., University of Delaware, 1998), 44–116. I am grateful to Bernard L. Herman, Nancy Van Dolsen, Dell Upton, Orlando Ridout V,

Joseph S. Wood, Billy G. Smith, Garry Wheeler Stone, Liam O. Riordan, members of the History Department at James Madison University, and participants in a seminar of the McNeil Center for Early American Studies for their many helpful comments on earlier versions.

1. Johann David Schoepf, *Travels in the Confederation [1783–1784]*, Volume 1: *New Jersey, Pennsylvania, Maryland, Virginia*, trans. and ed. Alfred J. Morrison (Philadelphia: William J. Campbell, 1911), 125.

2. Schoepf, *Travels in the Confederation*, 108. Contemporaries such as Schoepf often used the term "German," even though Pennsylvania Germans were an extraordinarily diverse group. For the purposes of expediency in this chapter, the terms "German" and the more cumbersome "Pennsylvania German" are used interchangeably despite this tremendous diversity.

3. Wilbur Zelinsky, *Nation into State: The Shifting Symbolic Foundations of American Nationalism* (Chapel Hill: University of North Carolina Press, 1988). Michael Kammen, *Mystic Chords of Memory: The Transformation of Tradition in American Culture* (New York: Alfred A. Knopf, 1991). Bernard L. Herman, "The Model Farmer and the Organization of the Countryside," in Catherine E. Hutchins, ed., *Everyday Life in the Early Republic* (Winterthur, Del.: Henry Francis du Pont Winterthur Museum, 1994), 35–59.

4. Thomas Anburey, *Travels Through the Interior Parts of America: In a series of letters. By an officer.* (London: William Lane, 1789), 284.

5. Anthony P. Cohen, *The Symbolic Construction of Community* (London: Routledge, 1989), 12–13, 117–118. Zelinsky, *Nation into State*, 16–17. Kammen, *Mystic Chords of Memory*, 65.

6. Dell Upton, "The City as Material Culture," in Anne Elizabeth Yentsch and Mary C. Beaudry, eds., *The Art and Mystery of Historical Archaeology: Essays in Honor of James Deetz* (Boca Raton La.: CRC Press, 1992), 51–53. See also Dell Upton, "Ethnicity, Authenticity, and Invented Traditions," *Historical Archaeology* 30, no. 2 (1996): 1–7.

7. James T. Lemon, "The Agricultural Practices of National Groups in Eighteenth-Century Southeastern Pennsylvania," *Geographical Review* 56, no. 4 (October 1966): 467–496. James T. Lemon, *The Best Poor Man's Country: A Geographical Study of Early Southeastern Pennsylvania* (New York: W. W. Norton, 1976). See also Russell Ferguson, "Introduction: Invisible Center," in Russell Ferguson, Martha Gever, Trinh T. Minh-ha, and Cornel West, eds., *Out There: Marginalization and Contemporary Cultures* (New York: New Museum of Contemporary Art and The MIT Press, 1990), 11, 12.

8. For more on the manner in which the tax was actually assessed, see Nancy Van Dolsen's essay in *The Domestic Landscape in the Early Republic: Architecture and Material Culture in Eighteenth-Century America*, Bernard L. Herman and Michael Steinitz, eds. (Knoxville: University of Tennessee Press, forthcoming). See also Lee Soltow, *Distribution of Wealth and Income in the United States in 1798* (Pittsburgh: University of Pittsburgh Press, 1989) and Michael Steinitz, "Rethinking Geographical Approaches to the Common House: The Evidence from Eighteenth-century Massachusetts," in Thomas Carter and Bernard L. Herman, eds., *Perspectives in Vernacular Architecture, III* (Columbia: University of Missouri Press, 1989), 16–26.

9. Schoepf, *Travels in the Confederation*, 103–104. Born in Germany in 1752,

Schoepf was thirty-two at the time he traveled through America. Although his studies in medicine and the natural sciences caused him to be especially sensitive to North American geology, climate, flora, and fauna, Schoepf was also an astute and thoughtful observer of humanity.

10. Rayner Wickersham Kelsey, ed., *Cazenove Journal 1794: A Record of the Journal of Theophile Cazenove through New Jersey and Pennsylvania* (Haverford: Pennsylvania History Press, 1922), 30, 69, 34. Cazenove, born of French parents in Amsterdam in 1740, was a financier and agent of the Holland Land Company. He became an American citizen in 1794 and was heavily involved in assessing land in western New York and Pennsylvania for speculative investment. Paul D. Evans, "Theophile Cazenove (October 13, 1740 to March 6, 1811)," in *Dictionary of American Biography* (1929), 580–581. Although Cazenove was often critical of the Germans' monetary practices, he defended their tendency to accumulate land. "They have all become rich, through the high price of grains since the French Revolution. They accumulate cash and keep it idle, by distrust—or they buy land, next to their own, which they do not cultivate and their savings remain idle. However, it is only fair to say that German farmers give farms to their sons as soon as they are of age, for their marriage, and even if they have 10 sons, they all become farmers,—while Irish farmers, if they make a fortune, bring up their children for the cities." Kelsey, *Cazenove Journal*, 44.

11. James Whitelaw, *Journal, 1773–93* (Manuscript B WS8J 6006, Vermont Historical Society), 31.

12. Lewis Evans, *A Brief Account of Pennsylvania* (1753). Quoted in Don Yoder, "Through the Traveler's Eye," in Alfred L. Shoemaker, ed., *The Pennsylvania Barn* (Lancaster, Pa.: Pennsylvania Dutch Folklore Center, 1955), 12–21, quote on 13.

13. François Alexandre Frédéric, duc de La Rochefoucauld-Liancourt, *Travels through the United States of North America, the country of the Iroquois, and upper Canada, in the years 1795, 1796, 1797; with an authentic account of Lower Canada*, 2 vols. (London: R. Phillips, 1799), 1:35, 46.

14. Benjamin Rush, "An Account of the Manners of the German Inhabitants of Pensylvania," *Columbian Magazine* (January 1789), 22–30. See also Benjamin Rush, *Essays Literary, Moral, and Philosophical*, ed. Michael Meranze (Schenectady, N.Y.: Union College Press, 1988) and Lemon, *Best Poor Man's Country*, xiv.

15. Tyrone Power, Esq., *Impressions of America; During the Years 1833, 1834, and 1835* (Philadelphia, 1836). Quoted in Yoder, "Through the Traveler's Eye," 12–21, 15.

16. Kathleen Neils Conzen, David A. Gerber, Ewa Morawska, George E. Pozzetta, and Rudolph J. Vecoli, "The Invention of Ethnicity: A Perspective from the U.S.A.," *Journal of American Ethnic History* 12, no. 1 (Fall 1992): 3–41, 4–5.

17. Conzen et al., "Invention of Ethnicity," 7–10. Dell Upton, "Ethnicity, Authenticity, and Invented Traditions," *Historical Archaeology* 30, no. 2 (1996): 1–7, 4.

18. Zelinsky, *Nation into State*, 17. Benedict R. Anderson, *Imagined Communities: Reflections on the Origin and Spread of Nationalism* (London: Verso, 1983, 1991, 1996).

19. La Rochefoucauld-Liancourt, *Travels through the United States*, 2:391–392.

20. Henry Bradshaw Fearon, *Sketches of America* (London: Strahan and Spottiswoode, 1819), 183–184.

21. Ferdinand-M. Bayard, *Travels of a Frenchman in Maryland and Virginia with a description of Philadelphia and Baltimore in 1791 or Travels in the Interior of the United States, to Bath, Winchester, in the Valley of the Shenandoah, etc., etc., during the Summer of 1791,* ed. and trans. Ben C. McCary (Ann Arbor, Mich.: Edwards Brothers, 1950), 98. Here the editor notes that Bayard is quoting from Jedidiah Morse (1761–1821), who is quoting in turn from Peter Kalm. Morse was the author of *Geography Made Easy* (1785), which was the first geography published in the U.S.

22. Kelsey, *Cazenove Journal,* 43–44.

23. Kelsey, *Cazenove Journal,* 34.

24. Francis Baily, *Journal of a Tour in Unsettled Parts of North America in 1796 & 1797* (London: Baily Brothers, 1856), 136–137.

25. Conzen et al., *Invention of Ethnicity,* 7–10.

26. Isaac Weld, *Travels Through the States of North America,* (London: William L. Clements, 1800), 125–127.

27. Weld, *Travels Through the States,* 122–124.

28. Kelsey, *Cazenove Journal,* 82–85.

29. Marianne S. Wokeck, "Patterns of German Settlements in the North American Colonies" (paper presented to the Philadelphia [now McNeil] Center for Early American Studies, November 4, 1994), 6–9, 16–17. Scott T. Swank, "The Germanic Fragment," in Scott T. Swank et al., *Arts of the Pennsylvania Germans* (New York: W. W. Norton, 1983), 5. Quote from Marianne S. Wokeck, *Trade in Strangers: The Beginnings of Mass Migration to North America* (University Park: Pennsylvania State University Press, 1999), xxviii.

30. Swank notes that most German immigrants followed the paths of total assimilation or controlled acculturation, and comparatively few completely rejected the mainstream culture. Swank, "The Germanic Fragment," 5.

31. See Philip E. Pendleton, *Oley Valley Heritage: The Colonial Years: 1700–1775* (Birdsboro, Pa.: Pennsylvania German Society, and Oley, Pa.: Oley Valley Heritage Association, 1994), 72–73. Upton, "Ethnicity, Authenticity, and Invented Traditions," 1–7.

32. Upton, "Ethnicity, Authenticity, and Invented Traditions," 4–5.

33. Charles Joyner, *Down by the Riverside: A South Carolina Slave Community* (Urbana: University of Illinois Press, 1984), xix–xxii. Dell Hymes, *Foundations in Sociolinguistics: An Ethnographic Approach* (Philadelphia: University of Pennsylvania Press, 1974), 47–51. Dell Hymes, *Pidginization and Creolization of Languages* (London: Cambridge University Press, 1971), 84. Dell Upton, commentary on "Mapping Difference in Early National America: Negotiating Identity Through the Construction of Place," American Studies Association Annual Meeting, Nashville, Tennessee, October 29, 1994. Kathleen Neils Conzen, "Mainstreams and Side Channels: The Localization of Immigrant Cultures," *Journal of American Ethnic History* 11, no. 1 (Fall 1991): 5–20. Upton, "Ethnicity, Authenticity, and Invented Traditions," 1–7. Conzen et al., "Invention of Ethnicity," 32.

34. Conzen et al., "Invention of Ethnicity," 8.

35. Construction of the Lancaster Turnpike began in 1792. Writing in 1796 and

1797, Francis Baily described the Lancaster Turnpike as "a masterpiece of its kind." Baily, *Journal of a Tour of North America*, 107. Historic Preservation Trust of Lancaster County, *Our Present Past: An Update of "Lancaster's Heritage"* (Lancaster, Pa.: Historic Preservation Trust of Lancaster County, 1985), 3–5. City of Lancaster, Department of Housing and Community Development, *The Historic City of Lancaster: A Report on its Historic Resources* (Lancaster, Pa.: Historic Preservation Trust of Lancaster County, 1995), 13–14.

36. The borough of Lititz was governed as well as settled by Moravians. It was created in 1756 "to serve Moravian Church members in the colonies and immigrants from Europe who desired close spiritual supervision but were not willing or fitted to live the religious communal life called for in the institutional arrangements at Bethlehem and Nazareth, Pennsylvania. At Bethlehem and Nazareth, Moravian members were required to live in community houses and pool their labor for the good of the congregation. At Lititz, each family would be permitted to live in their own house, if desired, and each family head could run his own business. . . . Every aspect of village life, religious, social, and economic, was under the supervision and control of the Aufseher Collegium, the administrative committee of the congregation. . . . Families were allowed to live in their own private homes, unlike the community houses in Bethlehem and Nazareth. Residents of Lititz were required to sign town regulations that controlled the community life rigidly." Lititz was not opened to non-Moravian inhabitants until 1855. Mary T. Wiley, "National Register Nomination for the Lititz Moravian Historic District," Lititz Borough, Lancaster County, Pa., October 1985. For a discussion of the ways in which Moravian community leaders in a North Carolina community balanced the maintenance of group cohesion and internal social control with the sometimes conflicting need to seem a part of the larger Anglo-American world, see Brian W. Thomas, "Inclusion and Exclusion in the Moravian Settlement in North Carolina, 1770–1790," *Historical Archaeology* 28, no. 3 (1994), 15–29.

37. For this study, the Pennsylvania German populations were determined by selecting a subpopulation of property owners with obviously Germanic surnames. When names were ambiguous and might be placed in either a Pennsylvania German or non–Pennsylvania German category, they were designated non–Pennsylvania German.

38. Weld, *Travels Through the States*, 125–127. Similarly, traveling between 1793 and 1797, William Priest noted: "All the back parts of Pennsylvania were in general cleared, and settled by german, and irish emigrants; but the former are commonly more prosperous than their neighbours, whom they excel in sobriety and economy, and have also a much better understanding amongst themselves. An irish family often arrives, and purchases a plantation; which for some years brings them good crops, but for want of manure will in time be worn out (a very common case in America). When in this situation they offer it for sale, the adjacent german families club a sum of money, purchase the land, plough it well, and let it remain in this state for three or four years; they then place an emigrant family from their *own country* upon the farm, who, by indefatigable industry and manure, soon bring the land round, pay for the estate by installments, and live very comfortably. Some of the best plantations in Pennsylvania were originally left in this manner. The irish family go two or three hundred miles up the country,

where they can purchase as much land as they please, from sixpence to a dollar per acre: here they literally *break fresh ground,* and begin the world again." William Priest, *Travels in the United States of America, Commencing in the year 1793, and ending in 1797* (London: J. Johnson, 1802), 61–62.

39. Evans, *A Brief Account of Pennsylvania,* quoted in Yoder, "Through the Traveler's Eye," 12–21, 13.

40. Historic Preservation Trust, *Our Present Past,* 136, 362.

41. Historic Preservation Trust, *Our Present Past,* 81.

42. While most accounts describing the churchlike aspects of Pennsylvania German barns appeared in the first half of the nineteenth century, a reference to the disproportionate size of these buildings appeared as early as 1753, right around the height of German in-migration to Pennsylvania. Evans, *A Brief Account of Pennsylvania.* Bernhard, Duke of Saxe-Weimar Eisenach, *Travels Through North America, During the Years 1825 and 1826* (Philadelphia, 1828). Charles Hoffman, *A Winter in the West* (1835). Francis J. Grund, *The Americans in Their Moral, Social, and Political Relations* (London, 1837). Quoted in Yoder, "Through the Traveler's Eye," 12–21.

43. Gabrielle Milan Lanier, "Samuel Wilson's Working World: Builders and Building in Chester County, Pennsylvania, 1780–1827," (Master's thesis, University of Delaware, 1989), 9–10. Lemon, *Best Poor Man's Country,* 79, 81, 83.

44. Contemporary description of "English" landscapes is provided by Brissot de Warville, who was one of the few observers to comment at some length on the effect the Quaker population had had on Philadelphia and the surrounding countryside. He praised the Quaker farmers in much the same way that Rush and others lauded German agricultural practices, and wrote, "If you would quit the town [Philadelphia], and walk over the farms of the Quakers, you will discover a greater degree of neatness, order, and care, among these cultivators than among any other." Jacques Pierre Brissot de Warville, *New Travels in the United States of America. In Two Volumes.* (London: J. S. Jordan, 1794), 387–388.

45. Scott T. Swank, "Proxemic Patterns," in Scott T. Swank et al., *Arts of the Pennsylvania Germans* (New York: W. W. Norton, 1983), 43, 52–53. See also Carol Kessler, "Ten Tulpehocken Inventories: What Do They Reveal about a Pennsylvania German Community?" *Pennsylvania Folklife* 33, no. 2 (Winter 1973–74), 16–30.

46. 1815 Direct Tax, Warwick Township, Lancaster County, Pa. See also Thomas A. Lainhoff, "The Buildings of Lancaster County, 1815" (Master's thesis, Pennsylvania State University, 1981), 121–141.

47. 1815 Direct Tax, Warwick Township.

48. 1815 Direct Tax, Warwick Township.

49. Scott Swank argues that total assimilation also often meant Anglicizing surnames. He based his assessment of the size of the Pennsylvania German population on the presence of Germanic surnames. Swank, "The Germanic Fragment," 4. See also Lemon, *Best Poor Man's Country,* 73–77, regarding population movement in late eighteenth-century Pennsylvania.

50. Anburey, *Travels Through the Interior of America,* 279–280.

51. John Beale Bordley, *Essays and Notes on Husbandry and Rural Affairs* (Philadelphia, 1799), quoted in Yoder, "Through the Traveler's Eye," 19.

52. William Cobbett, *A Year's Residence in America* (Boston: Small, Maynard, 1922), 25.

53. Thomas A. Lainhoff, "The Buildings of Lancaster County, 1815" (Master's thesis, Pennsylvania State University, 1981), 103–104.

54. Robert St. George has noted a similar inverse relationship between the decreasing availability of land and the increasing quantity of artifacts in the dwellings of the seventeenth-century New England yeomanry. See Robert Blair St. George, "'Set Thine House in Order': The Domestication of the Yeomanry in Seventeenth-Century New England," in Dell Upton and John Michael Vlach, eds., *Common Places: Readings in American Vernacular Architecture* (Athens: University of Georgia Press, 1986), 360. Conzen et al., "Invention of Ethnicity," 12–13. Cohen, *Symbolic Construction of Community,* 44, 46.

55. Lemon, "Agricultural Practices," 493–496. In reference to travelers' lies, plagiarisms, and prejudices in general, see also Percy G. Adams, *Travelers and Travel Liars, 1660–1800* (Berkeley: University of California Press, 1962), 1–18, 142–161, 191.

56. Edward A. Chappell, "Acculturation in the Shenandoah Valley: Rhenish Houses of the Massanutten Settlement," in Dell Upton and John Michael Vlach, eds., *Common Places: Readings in American Vernacular Architecture* (Athens: University of Georgia Press, 1986), 27–57.

57. Conzen et al., "Invention of Ethnicity," 12–13, 31.

58. Cohen, *Symbolic Construction of Community,* 15, 44, 46, 91, 102. Gerald L. Pocius, *A Place to Belong: Community Order and Everyday Space in Calvert, Newfoundland* (Athens: University of Georgia Press, and Montreal: McGill-Queen's University Press, 1991). Anthony P. Cohen, "Culture as Identity: An Anthropologist's View," *New Literary History* 24, no. 1 (Winter 1993): 195–209. Pressure toward this kind of "objectification" can result from processes other than acculturation alone. See Bernard. L. Herman, *The Stolen House* (Charlottesville: University Press of Virginia, 1992).

59. Swank, "The Germanic Fragment," 5.

60. Upton, "Mapping Difference."

61. Curiously, while this house dates to the eighteenth century, it does not appear on the 1798 tax rolls. One possible explanation is that the house stood close enough to the boundary line between Warwick and what was then Cocalico (now Ephrata) Township that it was assessed in the latter. Unfortunately, the 1798 tax records for Cocalico do not survive.

62. In 1798, houses owned by the wealthiest 10 percent of Warwick's population averaged 983 square feet on the first floor and 61 percent were built of stone, while dwellings owned by taxable residents who fell halfway down the scale—the fifth decile—averaged only 665 square feet and only 23 percent were constructed of stone. The original block of the Eby house contained 1,015 square feet of living space on the first floor.

63. Charles Lang Bergengren, "The Cycle of Transformations in the Houses of

Schaefferstown, Pennsylvania" (Ph.D. diss., University of Pennsylvania, 1988), 31–40. Also see Charles Bergengren, "From Lovers to Murderers: The Etiquette of Entry and the Social Implications of House Form," *Winterthur Portfolio* 29, no. 1 (Winter 1994): 43–72.

64. William J. Murtagh, *Moravian Architecture and Town Planning: Bethlehem, Pennsylvania and Other Eighteenth-Century American Settlements* (Chapel Hill: University of North Carolina Press, 1967), 124–127.

65. Although the original main block of the house actually measures 33 feet 6 inches by 43 feet 9½ inches, tax assessors in 1798 and 1815 recorded the estimated dimensions of 30 feet by 40 feet.

66. Ceilings insulated with varying mixtures of packed clay, straw, and sometimes rubble stone that were held in place by wooden pales (riven strips of wood usually around ⅜″ to ½″ thick) were common in many mid-eighteenth-century dwellings in German-settled regions. In some buildings, this type of construction was nailless; the wooden pales were slid into long grooves cut into both sides of the floor joists. In the Reist house the pales are boards measuring around 1 foot wide held in place with sash-sawn wooden strips applied with wrought nails.

67. See Pennsylvania Historic Resources Survey Form 071-55-15, Lancaster County.

68. As in both the Friedrich and Reist houses, a vaulted cellar may have been located beneath the now-demolished kitchen wing that was recorded in the 1815 tax record.

69. Nancy Van Dolsen, *Cumberland County: An Architectural Survey* (Carlisle, Pa.: Cumberland County Historical Society, 1990), 13–19.

70. Plans for at least one similarly designed dwelling are housed at the Moravian Archives in Bethlehem, Pennsylvania. Cynthia Falk has recently researched several similar side-passage dwellings in the German-settled regions of northern Chester County, Pennsylvania. See Cynthia Gayle Falk, "Evidence of Ethnicity and Status in the Architectural Landscape of Eighteenth-Century Coventry Township, Chester County, Pennsylvania" (Master's thesis, University of Delaware, 1996).

71. William Woys Weaver, "The Pennsylvania German House: European Antecedents and New World Forms," *Winterthur Portfolio* 21, no. 4 (Winter 1986): 243–264. For a romanticized image of how the *stube* functioned as the center of private homes in Germany and as an important aspect of domesticity, see John Louis Krimmel's *Der Nachtkartz, 1818–20* and *Woman Spinning,* reproduced in Anneliese Harding, *John Louis Krimmel: Genre Artist of the Early Republic* (Winterthur, Del.: Henry Francis du Pont Winterthur Museum, 1994), 110.

72. Quoted in Weaver, "The Pennsylvania German House," 257.

73. Henry Glassie, "Eighteenth-Century Cultural Process in Delaware Valley Folk Building," in Dell Upton and John Michael Vlach, eds., *Common Places: Readings in American Vernacular Architecture* (Athens: University of Georgia Press, 1986), 394–425, 407.

74. Bergengren, "Cycles of Transformation," 80–83.

75. Upton has argued that responses to the urban landscape often represent a con-

flation of such intentional and unintentional cultural byproducts. Upton, "The City as Material Culture," 51–53.

## CHAPTER 3

For their many helpful comments on earlier and more abbreviated versions of this essay, I am indebted to Cynthia Falk, Richard Haan, Bernard Herman, Graham Hodges, Carter Hudgins, Patricia Keller, and Jeanne Halgren Kilde.

1. This account is based on a story that appeared in *The Delaware Gazette* of April 17, 1829. The same story was reprinted in its entirety in *Niles' Register* the following week. "A Horrible Development," *Niles' Weekly Register,* (Baltimore: Hezekiah Niles, April 25, 1829), 144.

2. "A Horrible Development," 144. Dick Carter, *The History of Sussex County* (Millsboro, Del.: Delmarva News and Delaware Coast Press, 1976), 25. John A. Munroe, *History of Delaware,* 2nd ed. (Newark: University of Delaware Press, 1984), 282–283. *Narrative and Confessions of Lucretia P. Cannon, Who was Tried, Convicted, and Sentenced to be Hung at Georgetown, Delaware, with Two of Her Accomplices. Containing an Account of Some of the Most Horrible And Shocking Murders and Daring Robberies Ever Committed by One of the Female Sex.* (New York: New York Publishers, 1841), 9–19.

3. Carter, *History of Sussex County,* 25.

4. Quote from "A Horrible Development," 144. See also Enoch Lewis, ed., *The African Observer: A Monthly Journal Containing Essays and Documents Illustrative of the General Character and Moral and Political Effects, of Negro Slavery.* (Westport, Conn.: Negro Universities Press, 1970), 45; and Jesse Torrey, *American Slave Trade* (London: J. M. Cobbett, 1822), 74–76, 76n. Ira Berlin, *Slaves Without Masters: The Free Negro in the Antebellum South* (New York: Pantheon Books, 1974), 99–101, 160–161, 309; Carol Wilson, *Freedom at Risk: The Kidnapping of Free Blacks in America, 1780–1865* (Lexington: University Press of Kentucky, 1994), 18–37.

5. Bernard L. Herman, *The Stolen House* (Charlottesville: University Press of Virginia, 1992), 18.

6. *Narrative and Confessions of Lucretia P. Cannon,* 9–19.

7. Carter, *History of Sussex County,* 26.

8. Barre Toelken, *The Dynamics of Folklore,* rev. and expanded ed. (Logan: Utah State University Press, 1996), 403–406, 409–410. Robert A. Georges and Michael Owen Jones, *Folkloristics: An Introduction* (Bloomington: Indiana University Press, 1995), 84–86.

9. Robert Darnton, "Peasants Tell Tales: The Meaning of Mother Goose," in Robert Darnton, *The Great Cat Massacre and Other Episodes in French Cultural History* (New York: Vintage Books, 1985), 9–72, 53, 64.

10. William Cronon, "A Place for Stories: Nature, History, and Narrative," *Journal of American History* 78, no. 4 (March 1992), 1347–1376, 1350, 1354, 1367. See also Louis O. Mink, "Narrative Form as a Cognitive Instrument," in Robert H. Canary and Henry Kozicki, eds., *The Writing of History: Literary Form and Historical Understanding*

(Madison: University of Wisconsin Press, 1978), 129–149; David Demeritt, "Ecology, Objectivity and Critique in Writings on Nature and Human Societies," *Journal of Historical Geography* 20, no. 1 (January 1994), 22–37; William Cronon, "Comment: Cutting Loose or Running Aground?" *Journal of Historical Geography* 20, no. 1 (January 1994), 38–43.

11. Darnton, "Peasants Tell Tales," 9–72.

12. Patience Essah, *A House Divided: Slavery and Emancipation in Delaware, 1638–1865* (Charlottesville: University Press of Virginia, 1996), 10, 22. Present-day inhabitants of the state still readily acknowledge the presence of this cultural divide, although most would agree that the real dividing line between northern and southern Delaware is somewhere around the Chesapeake and Delaware Canal.

13. Jack Michel et al., *A Place in Time: Continuity and Change in Mid-Nineteenth-Century Delaware* (exhibit script) (Newark, Del.: Center for Historic Architecture and Engineering, 1985).

14. Bernard L. Herman, "Delaware's Orphans Court Valuations and the Reconstitution of Historic Landscapes, 1785–1830," in Peter Benes, ed., *Early American Probate Inventories* (Boston: Trustees of Boston University, 1989), 127–128. Cary Carson, Norman F. Barka, William M. Kelso, Garry Wheeler Stone, and Dell Upton, "Impermanent Architecture in the Southern American Colonies," in Robert Blair St. George, ed., *Material Life in America, 1600–1860* (Boston: Northeastern University Press, 1988).

15. Quoted in John A. Munroe, *Federalist Delaware 1775–1815* (New Brunswick, N.J.: Rutgers University Press and University of Delaware Press, 1954, 1987), 87. *Journals of Congress,* VII (1777–1788), 275.

16. Many of the disputed tracts in the North West Fork Hundred vicinity which fell within this zone were resurveyed by Thomas White and Rhoad Shankland. Shankland surveys, Sussex County Recorder of Deeds, Surveys A (1776), and Surveys B (1776), DSA. Munroe, *Federalist Delaware,* 16–17. George Read, ed., *Laws of the State of Delaware* (New Castle, Del.: printed by Samuel and John Adams, 1797–1816), 1:567–571. Henry C. Conrad, *History of the State of Delaware, From the Earliest Settlements to the Year 1907. Volume II.* (Wilmington, Del.: Wickersham, 1908), 683–684. Herman, *The Stolen House,* 31, 78. J. Thomas Scharf, *History of Delaware, 1609–1888* (Philadelphia: L. J. Richards and Company, 1888), 120–124, 1207. See also Herman, "Delaware's Orphans Court Valuations," 127–128.

17. Shankland surveys, A and B.

18. Carter, *History of Sussex County,* 48.

19. See T. H. Breen and Stephen Innes, *"Myne Own Ground": Race and Freedom on Virginia's Eastern Shore, 1640–1676* (New York: Oxford University Press, 1980), 40–41. See also Herman, "Delaware's Orphans Court Valuations," 127–128.

20. Barbara Jeanne Fields, *Slavery and Freedom on the Middle Ground: Maryland during the Nineteenth Century* (New Haven: Yale University Press, 1985), 3, 7–8, 19, 24, 36, 91.

21. As quoted in Essah, *A House Divided,* 5. From William K. Scarborough, ed., *The Diary of Edmund Ruffin,* 3 vols. (Baton Rouge: Louisiana State University Press, 1972), 1:630.

22. Essah, *A House Divided*, 22, 64.

23. Essah, *A House Divided*, 83–85.

24. William Currie, *An Historical Account of the Climate and Diseases of the United States of America* (Philadelphia: Dobson, 1792), 210–211. William Moraley's account of Delaware also referred to the state's low-lying areas as unhealthy, characterized by "swampy Ground, and Fevers raging continually." William Moraley, *The Infortunate: The Voyage and Adventures of William Moraley, an Indentured Servant*, ed. Susan E. Klepp and Billy G. Smith (University Park: Pennsylvania State University Press, 1992), 127. See also William Currie, "Face of the Country: Soil, Climate, and Productions of Delaware," *Delaware Register and Farmers' Magazine* 1, no. 2 (1838), 110–112.

25. François Alexandre Frédéric, duc de la Rochefoucauld-Liancourt, *Travels through the United States of North America, the country of the Iroquois, and upper Canada, in the years 1795, 1796, 1797; with an authentic account of Lower Canada*, 2 vols. (London: R. Phillips, 1799), 2:265–266.

26. La Rochefoucauld-Liancourt, *Travels through the United States*, 2:265–266.

27. Charles Wright, "Near Seaford, Sussex County, Delaware, December 3, 1851." Patent Office Report, 1852, Vol. 2: H. Document #102, 263. Bernard L. Herman explores in detail the dilemma posed by the wasteful and exploitative practices of timber cutting in the Sussex County forest districts in *The Stolen House*, 55, 64–75.

28. Frederick Douglass, *My Bondage and My Freedom* (Urbana: University of Illinois Press, 1987), 27.

29. Douglass, *My Bondage and My Freedom*, 44. Fields, *Slavery and Freedom*, 23.

30. George Alfred Townsend, *The Entailed Hat or Patty Cannon's Times* (New York: Harper & Brothers, 1884), 298.

31. Townsend, *The Entailed Hat*, 266–267. Harold Bell Hancock, *The Loyalists of Revolutionary Delaware* (Newark: University of Delaware Press, 1977).

32. Messenger, R. W., *Patty Cannon Administers Justice* (Cambridge, Md.: Tidewater, 1926, 1960), 69, 76n, 84.

33. Messenger, *Patty Cannon Administers Justice*, 69, 76n, 84.

34. Anthony P. Cohen, *The Symbolic Construction of Community* (London: Routledge, 1989), 12.

35. Kent C. Ryden, *Mapping the Invisible Landscape: Folklore, Writing, and the Sense of Place* (Iowa City: University of Iowa Press, 1993), 61.

36. Mary T. Hufford, "Telling the Landscape: Folklife Expressions and Sense of Place," in Rita Zorn Moonsammy, David Steven Cohen, and Lorraine E. Williams, eds., *Pinelands Folklife* (New Brunswick, N.J.: Rutgers University Press, 1987), 16, 23.

37. Underscoring the close architectural ties between North West Fork Hundred and Maryland, Henry Chandlee Forman has termed the Maston House "a Maryland House in Delaware." Henry Chandlee Forman, *Tidewater Maryland Architecture and Gardens* (New York: Architectural Book Publishing, 1956).

38. 1801, 1803, and 1816 Tax Assessments, North West Fork Hundred, Sussex County, Delaware, DSA.

39. Scharf, *History of Delaware*, 1305.

40. *William Morgan's Diary and Autobiography*, 218–219, DSA.

41. Most of the properties which the Cannon brothers were assessed for contained a dwelling. These parcels averaged around 175 acres apiece, and most consisted of around two-thirds improved land and one-third woodland. Typical examples include the "two hundred and sixty acres of land in ten. [tenancy] of Nath'l Patchett 130 improved with a single story framed dwelling 130 laid to be Wood land" and "one house and lot on the Road to Ferry." 1816 Tax Assessment, North West Fork Hundred, Sussex County, Delaware, DSA. Orphans court records indicate that the Cannons also owned property in adjacent hundreds.

42. *William Morgan's Diary and Autobiography,* 222–223. Quoted in Harold B. Hancock, ed., "William Morgan's Autobiography and Diary: Life in Sussex County, 1780–1857." *Delaware History* 19, no. 2 (Fall–Winter, 1980), 106–126. Interestingly, William Morgan rented the only house in the town of Cannon's Ferry that was *not* owned by the Cannon brothers.

43. See Herman, *The Stolen House,* 75–78.

44. Morgan Williams, 1795, G-50; Curtis Jacobs, 1831, P-384, both in SCOC.

45. When orphans court valuations for North West Fork Hundred taken between 1770 and 1830 are linked with contemporaneous tax lists, for example, the vast majority of cases listed fall in the top two deciles of the population in terms of taxable wealth and there are no valuations that represent individuals in the bottom 60 percent of the population. See also Herman, "Delaware's Orphans Court Valuations," 122–125.

46. William Bradley, 1808, K-183, K-184, SCOC.

47. William Bradley, 1808, K-183, K-184, SCOC. Home textile production was one of the most common household trades in Sussex County in the early nineteenth century, and the majority of surviving household inventories from North West Fork Hundred during this period listed a loom, a spinning wheel or two, and quantities of woolen and cotton thread. Looms and significant quantities of textile production equipment were also often associated with the presence of chattel labor. Nearly half of the households enumerated on the 1810 Delaware census owned looms; of those that did, more than one-third had slaves. William Bradley Probate Inventory, 1806, DSA.

48. Philip Hughes, 1808, K-129, SCOC.

49. When he discovered the bones, the farmer was renting land that belonged to Patty Cannon, and he probably occupied the house that stood on that property. A 1947 photograph of the purportedly original Cannon house taken just prior to its demolition shows what appears to be an earlier one-room, one-and-one-half-story gable-roofed frame house with several later additions appended to one gable. "Original Patty Cannon House Photo Revives Interest in Joe Johnson Disappearance" *Seaford Leader* (October 17, 1968), 68.

50. Henry Richards, 1791, F-197; John Elliott, 1811, L-52; Manlove Adams, 1814, L-275, all in SCOC. The most common house size in Sussex County in the late eighteenth and early nineteenth centuries was 18 by 20 feet, and at least 85 percent of all of the dwellings in the area consisted of one-story frame or log buildings measuring less than 20 by 25 feet. The smallest common house size was 12 by 12 feet. Herman,

*The Stolen House,* 183. United States Census, Delaware Manuscript Returns, North West Fork Hundred, 1810.

51. An example of ornamental brickwork appears faintly on the gable end wall of the Maston House (Figure 3.2). Uriah Brookfield, 1790, E-86; Elijah Kinney, 1810, K-408, both in SCOC. As in southwestern New Jersey, some houses began as small, probably one-room buildings that expanded by accretion. Orphans court assessments, which often enumerate the same dwelling at several points in time, occasionally capture the processes of physical deterioration, improvement, and architectural change. For example, in 1789, orphans court assessors described Stephen Coulbourn's weatherboarded log house and kitchen as two separate buildings connected by a partially enclosed passage, "a log dwelling house sixteen feet square with a brick chimney covered with clapboards in midling good repair, a log kitchen eighteen feet square with a brick chimney covered with plank in midling repair, & a passage between the two houses 6 feet wide covered with clapboards, in midling good repair." By 1793, the passage between the two buildings had apparently been fully enclosed; in that year, assessors recorded a slightly lengthened and improved "log dwelling house with an inside brick chimney 22 by 16 feet in good repair." Similarly, when court appointees assessed Levin Ricards' North West Fork Hundred mansion house in 1814 on behalf of his six surviving children, they described a 28-by-16-foot brick dwelling "in tolerable repair." Fourteen years later, the same brick building had been expanded with the addition of a frame wing attached to one gable and was described as "a mansion house forty six feet by sixteen one story, twenty six feet thereof brick, in bad repair the remaining twenty feet, a frame building in Middling repair." Stephen Coulbourn, 1789, 1793, E-16, E-17; Levin Ricards, 1814, 1828, L-248, O-263, both in SCOC.

52. Because some of the earliest settled parts of the Maryland and Virginia Eastern Shore constituted an important source area for subsequent migrations to Sussex County, many of the earliest architectural traditions that took root in this area related to the Chesapeake building traditions that prevailed in those areas. The appearance in Sussex County records—even now—of family names found also in Maryland and Virginia, demonstrates the strong links between this part of Delaware and the Eastern Shore source areas of Maryland and Virginia. Breen and Innes, *"Myne Owne Ground,"* 108. Herman, *The Stolen House,* 182. Carter, *History of Sussex County,* 9. United States Census, Delaware Manuscript Returns, Sussex County, 1800, 1810, 1820.

53. The entirety of Sussex County Orphans Court records contain only two other references to frame houses with brick ends, further underscoring their rarity in general and the significance of this small concentration in North West Fork. One was in Broadkiln and the other in Baltimore Hundred. Trustin Laws Polk, 1796, F-340; David Polk, 1776, A-376; Constantine Cannon, 1795, E-14, F-133, F-135; Levin Tull, 1792, G-5, all in SCOC. Herman, *The Stolen House,* 207.

54. An elite class of farmers who became increasingly removed from the real labor of farming arose in Delaware in the third quarter of the eighteenth century. These men devoted the bulk of their energies to amassing and managing agricultural estates. Like Philip Hughes, members of this rural elite who lived in the northern part of the state

often occupied large brick dwellings and were typically the wealthiest individuals in their communities. Their personal possessions also tended to include books, clocks, luxury items, and progressive agricultural implements such as wheat fans and corn shellers. Bernard L. Herman, Gabrielle M. Lanier, Rebecca J. Siders and Max Van Balgooy, *National Register of Historic Places: Dwellings of the Rural Elite in Central Delaware, 1770–1830 + / −.* (Newark, Del.: Center for Historic Architecture and Engineering, College of Urban Affairs and Public Policy, December 1989), E-1–E-4. Saxagotha Laws, 1797, F-315; Henry Hooper, 1797, G-11; Philip Hughes, 1804, IJ-327, all in SCOC. Henry Hooper Probate Inventory, 1796; Philip Hughes Probate Inventory, 1804, both in DSA.

55. Herman, "Delaware's Orphans Court Valuations," 131–132.

56. William Polk, 1813, L-202; Joshua Hays, 1829, P-89; Peter Waples, 1795, F-215; Obediah Smith, 1797, G-2; John Shepherd, 1784, D-315; Allin Short, 1794, E-99; Uriah Brookfield, 1790, E-86, all in SCOC. Herman, "Delaware's Orphans Court Valuations," 131–133.

57. SCOC, Quantification undertaken by the Center for Historic Architecture and Engineering, University of Delaware. Herman, *The Stolen House,* 203.

58. Henry Hooper, 1797, G-11; Hughitt Layton, 1812, L-88; David McElvane, 1774, A-253, all in SCOC.

59. Christopher Nutter, 1776, A-414; Trustin Laws Polk, 1796, F-340; Isaac Collins, 1784, D-239, all in SCOC. Room-by-room inventories often reveal the typical contents of domestic outbuildings such as smokehouses. For example, in 1790, Matthew Wilson's smokehouse contained "a parcel of Old Barrels & c," "about 10 lb of Bacon," and "about 50 lb of Dry'd Beef." Matthew Wilson Probate Inventory, 1790, Sussex County, Delaware, DSA.

60. Later-nineteenth-century milkhouses found on the Eastern Shore of Virginia were often small structures measuring less than four feet on a side. Often mounted on four poles or legs, these tiny buildings were readily movable and were typically placed against the north wall of the house or kitchen. Gabrielle M. Lanier and Bernard L. Herman, *Everyday Architecture of the Mid-Atlantic: Looking at Buildings and Landscapes* (Baltimore: Johns Hopkins University Press, 1997), 55. Carson et al., "Impermanent Architecture." Stephen Coulbourn, 1789, E-16; Jonathan Jacobs, 1808, K-147; Nathaniel Horsey, 1801, H-311; Whittington Williams, 1807, K-84, all in SCOC. David Richards Probate Inventory, 1797, DSA.

61. Nathaniel Horsey, 1794, E-39; Obediah Smith, 1797, G-2; Esther Lednum, 1828, O-261; Uriah Brookfield, 1790, E-86, all in SCOC. Herman, *The Stolen House,* 106–107.

62. Only 16 percent of all North West Fork properties listed in orphans court assessments between 1780 and 1830 included a barn. This percentage holds throughout all of Sussex County. Barns were often enlarged with sheds or lean-tos, as was White Jones's "framed Barn with two Sheds a Carriage house and stable the who[le] 30 feet by 16," and David Richards's "log Barn 20 feet by 18 in Bad repair with two Shades [sheds] in midling repair." White Jones, 1814, L-253; David Richards, 1798, G-14, both in SCOC. Herman, *The Stolen House,* 107.

63. Jonathan Jacobs, 1808, 1815, K-147, L-307; Alexander Laws, 1793, F-122; Thomas Hinds, 1774, A-245; William Hazzard, 1818, L-515, all in SCOC. Bernard Herman has found that most of the few surviving Sussex County barns tend to be small, timber-framed buildings supported by masonry piers and furnished with wooden floors. Herman, *The Stolen House*, 107–110. The barnyard often appeared as an extension of the barn, as inventory takers typically lumped objects found in the yard around the house with goods stored in the barn. Consequently, it is often difficult to determine with any certainty how farmers actually used their barns and what they stored there. Nevertheless, scattered inventory references suggest the way these buildings functioned. Thomas Sorden's farm in North West Fork included an old 32-by-30-foot barn filled with scythes and cradles, "1 old Dutch Fan," stored straw, wheat, and corn, several plows, harrows, hoes, and spades, axes, wedges, a horse-cart, tackle, cart wheels, and one "2 horse waggon & harness." A stack of cypress shingles and fourteen stacks of fodder appeared elsewhere in the inventory, and probably stood in a different building or outside in the yard. Because Sorden's inventory was taken in June, most of his crops, such as his "30 Bushels of Wheat seeded" were already growing in the ground. But when inventory takers inspected the contents of Saxagotha Laws's "sorry barn in bad repair" at the opposite end of the farmer's year, they found a different set of circumstances. Laws's barn contained farm-related tools such as plows, harrows, wedges, axes, a froe, sheep shears, and "1 old Dutch wheat fan," but it also sheltered quantities of fodder and stored crops, including beans, wheat, rye, and flax in the straw. Thomas Sorden Probate Inventory, June 5, 1812. Saxagotha Laws Probate Inventory, January 21, 1796, DSA. Herman, *The Stolen House*, 108, 117–118.

64. Hudson Cannon, 1811, L-47; William Hazzard, 1818, L-515, both in SCOC. David Morris's stable may have been earthfast, since it was described as "one Stable framed Twenty nine by ten in good repair, with posts in the ground." David Morris, 1829, P-85, SCOC. Ownership of horses and most other types of livestock declined in North West Fork between 1801 and 1816. In 1816, of a total of 670 taxable residents, just 304 owned a total of 662 horses. Of those horse owners, one individual owned 10, 17 were assessed for between 5 and 9, 71 owned 3 or 4, 112 owned just 2, and 103 kept a single horse. Because the Orphans Court Extracts cover a fifty-year span and the 1816 tax list captures a single year, it is difficult to make any meaningful comparisons between the number of stables enumerated and the number of horse owners present in the population at a given point in time. Still, the records suggest that horses may have been housed in other farm buildings or even out in the yard rather than in buildings that were stables alone. 1816 Tax Assessment, North West Fork Hundred, Sussex County, Delaware, DSA. H. John Michel, Jr., "A Typology of Delaware Farms, 1850" (paper delivered at the Organization of American Historians annual meeting, Los Angeles, April 1984). See also Herman, *The Stolen House*, 111.

65. Pennewell's tannery was one of only a very few listed in all of Sussex County. David Pennewell, 1824, 1829, N-196, O-454, SCOC.

66. William Polk, 1813, L-202; Obediah Smith, 1797, G-2; Henry Richards, 1791, F-197; Charles Rickards, 1794, F-84; Trustin Laws Polk, 1796, F-340, all in SCOC.

67. Orphans court assessors differentiated between the "Nine hundred and Thirty

one pannel of fence in midling repair" and the "pal'd Garden and yard" on Phillip Hughes's 109-acre property. Similarly, the property that Waitman Jones rented from Joseph Wyatt contained several different types of fencing, described as "three large Gatez in Good repair six set of Barz in Midling repair 1370 pannelz of fence in Good repair 93 ditto of cross fence in Midling repair 136 do of Division fence in bad repair 65 Do of straight fence around the pound Garden and Yard 57 ditto in Stack [stock] Penz." Nathaniel Horsey, 1791, E-10; Charles Dean, 1822, M-269; Levi Framton, 1810, K-334; Phillip Hughes, 1804, IJ-327; Joseph Wyatt, 1821, M-162, all in SCOC.

68. Robert Williams Probate Inventory, 1825; David Pennewell Probate Inventory, 1833; David Pennewell Coroner's Report, 1831, all in DSA.

69. In 1816, 45 percent of the population owned no livestock at all, 14 percent owned only 1 or 2 animals, 6 percent owned between 3 and 6, 6 percent owned between 7 and 12, 10 percent owned between 12 and 20, and 18 percent owned more than 21 animals. The typical livestock "starter kit" consisted of one cow or one horse or one of each. Those individuals who kept two animals were most likely to keep one cow and one horse. Those who owned three farm animals typically kept one more cow rather than one more horse. Farmers who owned between three and six animals tended to expand their holdings by keeping a few pigs and/or an occasional ox team. Individuals who owned between seven and twelve animals still typically owned no more than one or two horses; instead, they increased the number of pigs they kept, and also sometimes increased the number of oxen and cows. Individuals in this group also began to keep sheep. Only those individuals who owned more than a dozen animals tended to keep sheep with any regularity, however. Farmers who kept large numbers of livestock still tended to keep relatively small numbers of horses. 1816 Tax Assessments, North West Fork Hundred, Sussex County, Delaware, DSA.

70. Livestock, of course, constituted property first and foremost, and tax assessors and inventory takers calculated the value of most farm animals in monetary terms, listing the estimated value of each animal, often a description of its coloring if it was a large, valuable creature such as a cow or a horse, and sometimes its age. Still, many inventory takers also took special pains to include the names of the decedent's horses, suggesting a great deal about the level of affection and respect that often existed between farmers and some of their four-legged colleagues. When inventory takers assessed David Richards's estate in 1797, they not only described the coloring and markings of all of his cows, such as "1 deep Red Hiefer," "1 Wh$^t$ Back'd d$^o$," "1 Brindled Py'd Cow & Calf," but they also carefully listed the names of Richards's thirteen horses and two colts, noting a trio of mares named Poll, Fany, and Pleasure, a Sorrel mare named Kate, a stud horse named Partner, and a brown horse named Shakespear. And when Trustin Laws Polk's property was inventoried in 1796, enumerators carefully listed a black horse named Jack, a gray colt named Tomson, and a Red Cow and yearling named Love. Manlove Adams Probate Inventory, 1806–1807; David Richards Probate Inventory, 1797, both in DSA.

71. Bradley's household goods included three beds, several chests and chairs, a desk, two tables, and a loom and two spinning wheels. Most other households in the area reflected similar priorities: more than half of David Richards's probated wealth

consisted of livestock and transportation-related goods, and just under half of Manlove Adams's estate was tied up in farm equipment and livestock. Bradley's probate inventory breaks down as follows: 48 percent in slaves, 18 percent in livestock, 18 percent in farming and transportation-related goods, 11 percent in household furnishings, kitchenware, and household goods, 2 percent in food and clothing, and 3 percent in home textile production or other home manufactures. A sample of other probate inventories from North West Fork Hundred taken between 1780 and 1830 reveals a similar strategy of investment at the time the inventories were taken, with household goods and furnishings usually amounting to around 10 percent or so and the combination of slaves, livestock, and farm-related goods totaling between 78 and 84 percent of the total inventoried wealth. Sussex County Probate Inventories, DSA.

72. Essah, *A House Divided,* 83.

73. Will of Hughett Cannon, March 14, 1809. Hughett Cannon Probate Inventory, March 30, 1809, DSA.

74. Probate inventories not only recorded the name, age, and valuation of a decedent's slaves but often specified the terms of manumission. John Wilson's 1826 inventory lists fifteen slaves, among them "1 Negro Man Named Henry to Serve 10 Years." Similarly, Abidnigo Elliott died in 1825 with four slaves, each of whom carried a different term of service: the girl Mahala was enslaved for life, while her companions Ann and Catherine would be freed after at least another twenty years and a boy named Mitchel would gain freedom at the age of 25. United States Census, Delaware Manuscript Returns, Sussex County, 1800, 1810, 1820. Probate inventories of Clement Ross, 1808; Curtis Jacobs, 1831; John Wilson, 1826; and Abidnigo Elliott, 1825, all in DSA. Herman, *The Stolen House,* 34–36.

75. Herman, *The Stolen House,* 37. Levin Cannon, 1791, E-12; Trustin Laws Polk, 1796, F-340, both in SCOC. Levin Cannon Probate Inventory, 1789; Trustin Laws Polk Probate Inventory, 1796, both in DSA.

76. Documentary evidence for free black estates is sparse, but two Sussex County probate inventories from this period suggest the range of goods owned by some of the more economically secure members of this population. John Laws, described by appraisers as a "Free Negro" residing in North West Fork Hundred at his death in 1820, was assessed for a modest estate that included household furnishings, textile processing equipment, farm tools and equipment, and livestock such as mares, pigs, a milk cow, and geese. Like many other Sussex County farmers, he grew flax, wheat, oats, and corn and maintained an orchard. Robert Williams, also described as a free Negro, was assessed for a similar range of goods. His estate included the usual household goods and farm and textile processing equipment, but Williams also owned cooper's tools, dairying equipment, and livestock such as horses, cattle, roaming pigs ("1 Sow 6 pigs as they run"), and sheep. Based on the contents of his inventory, Williams produced oats, corn, and rye on his farm. John Laws Probate Inventory, 1820; Robert Williams Probate Inventory, both in DSA.

77. Lewis, *The African Observer,* 45.

78. Torrey, *American Slave Trade,* 99–101, 160–161, 309. Wilson, *Freedom at Risk,* 18–37.

79. Herman, *The Stolen House,* 182–183. New Castle and Kent County Orphans Court Extracts, 1770–1830. 1798 Direct Tax, Sadsbury Township, Chester County, Pennsylvania. 1798 Direct Tax, Hempfield, Warwick, and Conestoga Townships, Lancaster County, Pennsylvania. 1815 Direct Tax, Warwick Township, Lancaster County, Pennsylvania.

80. Herman, "Delaware's Orphans Court Valuations," 133–136.

81. A sample of Sussex County probate inventories taken between 1786 and 1825 shows most farmers growing corn, wheat, fruit, and flax, but many inventories also include large quantities of planking, tanbark, or shingles.

82. In 1816, tenants worked 41 percent of the land in North West Fork. 1816 Tax Assessments, North West Fork Hundred.

## CHAPTER 4

1. This and the other Thomas Yorke photographs in this work come from the Thomas Yorke Photograph Collection, 1888, SCHS.

2. Thomas Shourds, *History and Genealogy of Fenwick's Colony* (Bridgeton, N.J.: George F. Nixon, Publisher, 1876).

3. Personal communication with James Turk, director of the Salem County Historical Society, May 6, 1997. Turk maintains that the coincidence of the nation's centennial and Salem's bicentennial, which occurred within a year of each other and produced a flurry of interest in the region's early history, spawned not only the work of Yorke and Simkins but the founding of the Salem County Historical Society itself.

4. Joseph Sickler, *The Old Houses of Salem County,* 2nd ed. (Salem, N.J.: Sunbeam Publishing), 1949.

5. For a provocative discussion of mapping, see Svetlana Alpers, *The Art of Describing: Dutch Art in the Seventeenth Century* (Chicago: University of Chicago Press, 1983). Also related is Howard Morphy's work on the relationship of Australian Aboriginal people (Yolngu) to the landscape through the mediating process of the ancestral past. Morphy argues that, for the Yolngu, landscape is not just an "intervening sign system that serves the purpose of passing on information about the ancestral past" but is actually integral to the message. Howard Morphy, "Landscape and the Reproduction of the Ancestral Past," in Eric Hirsch and Michael O'Hanlon, eds., *The Anthropology of Landscape: Perspectives on Place and Space* (Oxford: Oxford University Press, 1995), 184–209. See also J. B. Harley, "Deconstructing the Map," in Trevor J. Barnes and James S. Duncan, eds., *Writing Worlds: Discourse, Text and Metaphor in the Representation of Landscape* (London: Routledge, 1992), 231–247. Cynthia Van Zandt, "Knowledge as a Source of Power in Cross-Cultural Encounters in Early Colonial North America and the Caribbean" (paper presented to the McNeil Center for Early American Studies, April 21, 1995, Princeton University). Regarding the lasting nature of the tradition of documenting the area's ancestral map, even the early work of the Historic American Buildings Survey in the 1930s followed this same path. The pattern-end houses of

southwestern New Jersey were typically documented in great detail, while less "significant" buildings often escaped notice.

6. Sickler, *Old Houses of Salem County*, 6.

7. In his work on the nature of architectural interpretation, Juan Pablo Bonta has argued that time and its passage can change or "wear down" the original meanings inherent in architecture. As buildings and other objects pass through stages of time and use, they gain and lose currency, often experiencing a process of resignification and entering an area of reified value. Juan Pablo Bonta, *Architecture and Its Interpretation: A Study of Expressive Systems in Architecture* (New York: Rizzoli, 1979), 175–224.

8. James Deetz, "Landscapes as Cultural Statements," in William M. Kelso and Rachel M. Most, eds., *Earth Patterns: Essays in Landscape Archaeology* (Charlottesville: University Press of Virginia, 1990), 1.

9. Fernand Braudel, *On History*, trans. Sarah Matthews (Chicago: University of Chicago Press, 1980), 25–54.

10. Denis Wood, with John Fels, *The Power of Maps* (New York: Guilford Press, 1992), 72.

11. J. B. Harley, "Maps, Knowledge, and Power," in Dennis Cosgrove and Stephen Daniels, eds., *The Iconography of Landscape: Essays on the Symbolic Representation, Design, and Use of Past Environments* (Cambridge: Cambridge University Press, 1988), 278. Harley, "Deconstructing the Map," 247. J. B. Harley and Paul Laxton, *The New Nature of Maps: Essays in the History of Cartography* (Baltimore: Johns Hopkins University Press, 2001).

12. Peter Hugill, "The Landscape as a Code for Conduct: Reflections on Its Role in Walter Firey's 'Aesthetic-Historical-Genealogical Complex,'" *Geoscience and Man* 24 (April 30, 1984): 21–30. Hugill based much of this article on concepts advanced in Walter Firey, *Land Use in Central Boston* (Cambridge: Harvard University Press, 1947) and Walter Firey, "Sentiment and Symbolism as Ecological Variables," *American Sociological Review* 10, no. 2 (April 1945): 140–148. For more on the ways in which elites both shape landscape and identify with that shaped landscape, see also Peter J. Hugill, "Houses in Cazenovia: The Effects of Time and Class" *Landscape* 24, no. 2 (1980): 10–15; James S. Duncan, Jr., "Landscape Taste as a Symbol of Group Identity: A Westchester County Village," *Geographical Review* 63, no. 3 (July 1973): 334–355; Peter J. Hugill, *Upstate Arcadia: Landscape, Aesthetics, and the Triumph of Social Differentiation in America* (Lanham, Md.: Rowman & Littlefield Publishers, 1995); Peter J. Hugill, "Home and Class among an American Landed Elite," in John A. Agnew and James S. Duncan, eds., *The Power of Place: Bringing Together Geographical and Sociological Imaginations* (Boston: Unwin Hyman, 1989): 66–80; Peter J. Hugill, "English Landscape Tastes in the United States," *Geographical Review* 76, no. 4 (October 1986): 408–423; and Peter J. Hugill, "Social Conduct on the Golden Mile," *Annals of the Association of American Geographers* 65, no. 2 (June 1975): 214–228. George E. Marcus, "'Elite' as a Concept, Theory, and Research Tradition," in George E. Marcus, ed., *Elites: Ethnographic Issues* (Albuquerque: School of American Research and University of New Mexico Press, 1983), 7–27.

13. Wilbur Zelinsky termed this phenomenon the Doctrine of First Effective Settlement. Zelinsky, *The Cultural Geography of the United States* (Englewood Cliffs, N.J.: Prentice-Hall, 1973), 13–14. Before Zelinsky, Fred Kniffen discussed the same phenomenon, which he termed "initial occupance." Fred B. Kniffen, "Folk Housing: Key to Diffusion," in Dell Upton and John Michael Vlach, eds., *Common Places: Readings in American Vernacular Architecture* (Athens: University of Georgia Press, 1986), 5–6, first published in *Annals of the Association of American Geographers* 55, no. 4 (December 1965): 549–577.

14. Peter Wacker has argued that New Jersey's settlement landscape has tended to follow in this tradition of first effective settlement or initial occupance, and he contends that preexisting aboriginal routes and settlements also impacted subsequent European choices and all choices that followed. Peter O. Wacker, *Land and People: A Cultural Geography of Preindustrial New Jersey: A Historical Geography* (New Brunswick, N.J.: Rutgers University Press, 1975), 412.

15. For more on Quaker tribal authority, see Susan S. Forbes, "Quaker Tribalism," in Michael Zuckerman, ed., *Friends and Neighbors: Group Life in America's First Plural Society* (Philadelphia: Temple University Press, 1982), 145–173.

16. Michael Chiarappa, "'The First and Best Sort': Quakerism, Brick Artisanry, and the Vernacular Aesthetics of Eighteenth-Century West New Jersey Pattern Brickwork Architecture (Ph.D. diss., University of Pennsylvania, 1992), 264–270. See also Robert David Sack, *Human Territoriality: Its Theory and History* (Cambridge: Cambridge University Press, 1986), Michael J. Chiarappa, "The Social Context of Eighteenth-Century West New Jersey Brick Artisanry," in Thomas Carter and Bernard L. Herman, eds., *Perspectives in Vernacular Architecture, IV* (Columbia: University of Missouri Press, 1991), 31–43; and Susan Garfinkel, "Genres of Worldliness: Meanings of the Meeting House for Philadelphia Friends, 1755–1830" (Ph.D. diss., University of Pennsylvania, 1997). For more on the linkages between houses and religious buildings, see Abbott Lowell Cummings, "Meeting and Dwelling House: Interrelationships in Early New England," in Peter Benes and Philip Zimmerman, eds., *New England Meeting House and Church: 1630–1850* (Boston, Mass., and Dublin, N.H.: Boston University and Dublin Seminar for New England Folklife, 1979) and Dell Upton, *Holy Things and Profane: Anglican Parish Churches in Colonial Virginia* (Cambridge and New York: MIT Press and Architectural History Foundation, 1986), 108–110, 160, 164. For more on the notion of a regional visual architectural community constructed by the interplay between sacred and secular forms on the landscape, see Gabrielle M. Lanier and Bernard L. Herman, *Everyday Architecture of the Mid-Atlantic: Looking at Buildings and Landscapes* (Baltimore: Johns Hopkins University Press, 1997), 269–274.

17. For more on the relationship between elites and the architectural landscape of eighteenth-century Massachusetts, see Kevin M. Sweeney, "Mansion People: Kinship, Class, and Architecture in Western Massachusetts in the Mid Eighteenth Century" *Winterthur Portfolio* 19 (Winter 1984): 231–255. Alan Gowans has argued that the distinctive patterned brick "mansions" of southwestern New Jersey were the monuments or "family institutions" of a Homeric aristocracy. Gowans, "The Mansions of Alloways Creek," in Dell Upton and John Michael Vlach, eds., *Common Places: Readings in Amer-*

*ican Vernacular Architecture* (Athens: University of Georgia Press, 1986), 367–393. For more on the Quaker influence in the Delaware Valley, see Barry Levy, *Quakers and the American Family: British Settlement in the Delaware Valley* (New York: Oxford University Press, 1988); James T. Lemon, *The Best Poor Man's Country: A Geographical Study of Early Southeastern Pennsylvania* (New York: W. W. Norton, 1976); and David Hackett Fischer, *Albion's Seed: Four British Folkways in America* (New York: Oxford University Press, 1989). For more on the Quaker influence in New Jersey, see John E. Pomfret, *The Province of West New Jersey, 1609–1702: A History of the Origins of an American Colony* (Princeton: Princeton University Press, 1956). For a discussion of the way status intersected with ethnicity in the built environment of early Pennsylvania, see Cynthia G. Falk, "Symbols of Assimilation or Status? The Meanings of Eighteenth-Century Houses in Coventry Township, Chester County, Pennsylvania," *Winterthur Portfolio* 33, no. 2–3 (1998), 107–134.

18. Thomas L. Purvis, "'High-Born, Long-Recorded Families': Social Origins of New Jersey Assemblymen, 1703 to 1776," *William and Mary Quarterly*, 3rd ser., vol. 37, no. 4 (October 1980): 592–615, quote on 615.

19. Chiarappa, "'The First and Best Sort,'" 282.

20. Peter O. Wacker and Paul G. E. Clemens, *Land Use in Early New Jersey: A Historical Geography* (Newark: New Jersey Historical Society, 1995), 120–121. Robert G. LeBlanc, "The Differential Perception of Salt Marshes by the Folk and Elite in the Late Nineteenth Century" (unpublished paper, 1978). Cited in Wacker and Clemens, *Land Use in Early New Jersey*, 52n. Elite landscape imprints do not always prevail uniformly, of course. In her study of farmhouses in Sutton, Massachusetts, Nora Pat Small showed how the middling rather than the wealthiest rural farmers transformed the built landscape with their classically ornamented two-story-with-ell dwellings. Small maintains that, when rural farmers changed their own houses to adapt to new living standards and livelihoods, a distant elite bent on retaining the rural virtue they had long associated with small farms and farm buildings unsuccessfully attempted to control that reordering by advocating reform. In this instance, elite reformers appeared to be protecting their own position by linking the concept of virtue to the small yeoman farmer, thus keeping middling farmers "in their place." When farmers of average means owned too much land or built houses with too many rooms, the reformers maintained, they created waste and disorder because they lacked the requisite resources to manage it all. Large farms and houses should remain the sole province of gentlemen who could properly afford to take care of them, they reasoned. Nora Pat Small, "The Search for a New Rural Order: Farmhouses in Sutton, Massachusetts, 1790–1830," *William and Mary Quarterly*, 3rd ser., vol. 53, no. 1 (January 1996): 67–86.

21. Michael Chiarappa has suggested that many who claimed London as their place of residence may have recently moved to that city from the north and west of England, and so the connection between the Salem County region and the northwestern section of England may be even stronger than records imply. Chiarappa, "'The First and Best Sort,'" 8, 54n. Levy, *Quakers and the American Family*, 6–15, 25–35. Fischer, *Albion's Seed*, 441–442.

22. As quoted in Joseph S. Sickler, *The History of Salem County, New Jersey, Being the*

*Story of John Fenwick's Colony, the Oldest English Speaking Settlement on the Delaware River* (Salem, N.J.: Sunbeam, 1937), 22–23.

23. As quoted in Kimberly R. Sebold, *From Marsh to Farm: The Landscape Transformation of Coastal New Jersey* (Washington, D.C.: Historic American Buildings Survey/Historic American Engineering Record/National Park Service, 1992), 29.

24. Accessible Archives, *Pennsylvania Gazette,* CD-ROM (Provo, Utah: Folio Corp., 1991), January 19, 1769.

25. Accessible Archives, *Pennsylvania Gazette,* CD-ROM (Provo, Utah: Folio Corp., 1991), January 26, 1769.

26. Chiarappa, "'The First and Best Sort,'" 76.

27. F. Alan Palmer, ed. *The Beloved Cohansie of Philip Vickers Fithian.* (Greenwich, N.J.: Cumberland County Historical Society, 1990), 198, 230–231. Fithian was once obliged to lodge overnight with a friend across the river in Port Penn when ferry travel to New Jersey had been temporarily interrupted. See also Philip Vickers Fithian, "Journal and Letters, 1766–1767," Special Collections, Firestone Library, Princeton University; and Philip Vickers Fithian, *Journal and Letters of Philip Vickers Fithian, 1773–1774,* ed. Hunter D. Farish (Williamsburg, Va.: Colonial Williamsburg, 1957). In 1817 a steamboat ran the 19-mile distance between Philadelphia and Burlington, New Jersey, which is upriver from Philadelphia and well upriver from Salem County, in 1 hour and 40 minutes. *Niles' Weekly Register,* June 38, 1817, 287. The *West Jersey Gazette* began to advertise steamboat and packet services on the Delaware fairly regularly around 1818. *West Jersey Gazette,* August 20, 1817 through September 20, 1819, microfilm, SCHS. By 1826, there were at least seventeen steamboats plying the Delaware River, and plans were in place to begin operating several more. *Niles' Weekly Register,* June 17, 1826, 283. On architectural commonalities between southwestern New Jersey and coastal Delaware, see Lanier and Herman, *Everyday Architecture.*

28. For a discussion of the way the organization of the Salem County countryside relates to other Delaware Valley landscapes, see Bernard L. Herman, "The Model Farmer and the Organization of the Countryside," in Catherine E. Hutchins, ed., *Everyday Life in the Early Republic* (Winterthur, Del.: Henry Francis du Pont Winterthur Museum, 1994), 35–59.

29. Tax rolls typically included five categories of taxable residents: "persons," "householders," "single men with horse," "single men," and "free negroes." Individuals who owned land fell in the category of "persons" and were listed first; these taxables were usually also assessed for livestock and occasionally for shares in other taxable items such as tanyards, grist mills, or vessels. Individuals who did not own land fell in one of the other four categories. In 1798, the population of Pittsgrove was about 31 percent landless, and landowners held an average of 73 improved acres; while in Mannington, which was 41 percent landless, landowners averaged about 125 improved acres. Jackson Turner Main, *The Social Structure of Revolutionary America* (Princeton: Princeton University Press, 1965), 25, 33. See also Gowans, "The Mansions of Alloways Creek," 391–392n.

30. Data for tenancy rates and landholding patterns in Mannington Township were compiled by cross-linking the "A" list of the 1798 direct tax—the only portion of the

list that survives for Mannington and Pittsgrove—with a local tax assessed in 1797. Although the direct tax counted only property owners, the 1797 tax included both propertied and propertiless residents in the township. Data for Pittsgrove Township was compiled in a similar fashion, by linking the "A" list of the 1798 direct tax to a 1798 local list, which also included landless residents. Data for Lower Alloways Creek was compiled by linking the "A" and "B" lists of the 1798 direct tax. Lower Alloways Creek is the only township in Salem County for which both the "A" and "B" lists of the direct tax survive. Tenancy rates in 1798 stood at 39 percent in Lower Alloways Creek and 37 percent in Pittsgrove. For more on tenancy in the Delaware Valley, see Lucy Simler, "Tenancy in Colonial Pennsylvania: The Case of Chester County," *William and Mary Quarterly* 43, no. 4 (October 1986): 542–569; Lemon, *Best Poor Man's Country*, 96–97; J. Ritchie Garrison, "Tenancy and Farming," in J. Ritchie Garrison, Bernard L. Herman, and Barbara McLean Ward, eds., *After Ratification: Material Life in Delaware, 1789–1820* (Newark, Del.: Museum Studies Program, 1988), 21–37.

31. In Lower Alloways Creek, more than one-third (39%) of all the enumerated properties were tenant-occupied in 1798. Comparison between owner and tenant populations in this township yields a series of interesting results. First, in socioeconomic terms, tenants were dispersed throughout all levels of the population. Second, comparative statistics for the mean acreage and total assessed property value for tenants and owner-occupants reveal almost no visible difference; in fact, tenants were likely to occupy properties that were valued slightly *higher* than those occupied by owner-occupants. Third, there was almost no difference in the amount of space contained in tenant-occupied and owner-occupied dwellings. Most ratable residents, whether they were owners or tenants, lived in houses averaging about 450 square feet on the first floor. Finally, similarities between owner and tenant populations extend even to the size of agricultural buildings. Barns were no more likely than dwellings to be larger if they stood on owner-occupied properties. Peter Wacker has also cited this similarity between owner and tenant populations in Salem County, noting that houses valued at more than $100 in New England–settled Monmouth County were almost always occupied by the owner, "while in Salem County almost as many such houses were occupied by tenants." Peter O. Wacker, "Relations Between Cultural Origins, Relative Wealth, and the Size, Form, and Materials of Construction of Rural Dwellings in New Jersey During the Eighteenth Century," in Ch. Higounet, ed., *Geographie Historique: Du Village et de la Maison Rurale* (Paris: Centre National de la Recherche Scientifique, 1979), 216.

32. For the sake of comparison, the following figures are provided. They are derived from the 1798 direct tax and local tax assessments taken in chronologically close years. In Warwick, Hempfield, and Conestoga, the three townships in Lancaster County, Pennsylvania, examined in Chapter 2, landholdings averaged between 64 and 103 acres in 1798. In Sadsbury Township on the western border of Chester County, Pennsylvania, the average ratable resident was assessed for 129 acres in the same year, while in Tredyffrin Township, located farther east in Chester County, the average landholding was 84 acres. Further to the south and just over a decade later, the average landholding in Baltimore Hundred in Sussex County, Delaware, was about 200 acres in 1809, although because land in this area was concentrated in the hands of a few

wealthy property owners, the *median* property size was only 120 acres. Information for Baltimore Hundred from Bernard L. Herman, *The Stolen House* (Charlottesville: University Press of Virginia, 1992), 96. 1798 federal direct tax for Warwick, Hempfield, and Conestoga Townships, Lancaster County, Pennsylvania, and for Sadsbury and Tredyffrin Townships, Chester County, Pennsylvania.

33. Wacker and Clemens, *Land Use in Early New Jersey,* 96, 257. See also Wacker, "Relations Between Cultural Origins," 215.

34. Wacker and Clemens, *Land Use in Early New Jersey,* 70–72, 85n.

35. Land in nearby Cumberland County experienced less improvement. According to Wacker's findings, Salem County was 61.5 percent improved by 1751, 75.6 percent improved in 1769, and 76.6 percent improved in 1784. In Cumberland County in the same years, the landscape was 36.7 percent, 53.4 percent, and 53.2 percent improved. Wacker and Clemens, *Land Use in Early New Jersey,* 70. Information on statewide and countywide soil improvement from Wacker and Clemens, *Land Use in Early New Jersey,* 66, 70–71, 73–74, 117–118. Information on improved land in Mannington Township from 1797 local tax, Mannington Township, Salem County, New Jersey. Wacker also found that prevailing eighteenth-century perceptions of soil fertility were generally accurate: based on a systematic evaluation of New Jersey lands for tax purposes in 1752, the inner coastal plain and the Piedmont regions were generally perceived to have the most fertile soils. Wacker and Clemens, *Land Use in Early New Jersey,* 117–120.

36. Palmer, *Beloved Cohansie of Philip Vickers Fithian,* 45. "Manuscript Diary of Richard Wood," *South Jersey Magazine* (Fall 1974–Fall 1984).

37. Accessible Archives, *Pennsylvania Gazette,* CD-ROM, August 6, 1767.

38. About 100 acres of this plantation consisted of cleared upland; the rest was "well-timbered." The upland acreage was reported as, "very strong" and capable of supporting good wheat, other grain, or grass. Also standing on the premises were a new two-story frame dwelling and a large log kitchen; the advertisement promised "a barn of 30 feet by 50, with stalls for 16 oxen, is intended to be built and finished by next harvest." The place was "well suited for a large dairy, or for feeding cattle, and excellent range for hogs." Accessible Archives, *Pennsylvania Gazette,* CD-ROM, January 26, 1769.

39. Sebold, *From Marsh to Farm,* 3. See also Donald Worster, *Rivers of Empire: Water, Aridity, and the Growth of the American West* (New York: Pantheon Books, 1985), 123–124.

40. Wacker and Clemens, *Land Use in Early New Jersey,* 121.

41. Wacker and Clemens, *Land Use in Early New Jersey,* 120. Adolph B. Benson, ed. and trans., *Peter Kalm's Travels in North America: The English Version of 1770* (New York: Wilson-Erickson, 1937) 1:181.

42. Palmer, *Beloved Cohansie of Philip Vickers Fithian,* 82–83, 95. "Manuscript Diary of Richard Wood," *South Jersey Magazine* 8, no. 1 (January–March 1979), 24.

43. Palmer, *Beloved Cohansie of Philip Vickers Fithian,* 89.

44. Accessible Archives, *Pennsylvania Gazette,* CD-ROM, March 22, 1770 and March 5, 1783.

45. Accessible Archives, *Pennsylvania Gazette,* CD-ROM, January 19, 1769.

46. Robert Gibbon Johnson, "On Reclaiming Marsh Land," *American Farmer* 8, no.

24 (September 1, 1826): 185–187. See also Robert Gibbon Johnson, "On Reclaiming Marsh Land," *American Farmer* 8, no. 25 (September 8, 1826): 193–194 and no. 26 (September 15, 1826): 201–202. For an equally detailed explanation of methods for embanking marsh land in the Carolina low country, see James C. Darby, "On the Embanking and Preparation of Marsh Land, for the Cultivation of Rice," *Southern Agriculturist, Horticulturist, and Register of Rural Affairs* 2 (January 1829): 23–28.

47. David J. Grettler, "The Landscape of Reform: Society, Environment, and Agricultural Reform in Central Delaware, 1790–1840" (Ph.D. diss., University of Delaware, 1990), 166–67. Sebold, *Marsh to Farm*, 21–28. Lanier and Herman, *Everyday Architecture*, 292–95.

48. Palmer, *Beloved Cohansie of Philip Vickers Fithian,* 31–32, 35.

49. "Manuscript Diary of Richard Wood," April 17, 1810 and August 24, 1810.

50. Robert G. Johnson, Esq., Account Book, 1786–1850, SCHS, MS #5.

51. Wacker and Clemens, *Land Use in Early New Jersey,* 120–121. Wacker bases part of his argument on an unpublished study by Robert G. LeBlanc, "Differential Perception of Salt Marshes" (Wacker and Clemens, *Land Use in Early New Jersey,* 52n). For more on the impact of a rural agricultural elite on the built environment during the same time period but on the opposite side of the Delaware River, see Bernard L. Herman, Gabrielle M. Lanier, Rebecca J. Siders and Max Van Balgooy, "National Register of Historic Places: Dwellings of the Rural Elite in Central Delaware, 1770–1830 +/−" (Newark: Center for Historic Architecture and Engineering, College of Urban Affairs and Public Policy, University of Delaware, 1989), E1–E18. In his study of the landscape and attitudinal changes wrought by agricultural reform in Kent County, Delaware, David J. Grettler has also linked marsh reclamation with a landed elite. In Delaware, Grettler found that marsh improvement projects tended to be initiated by wealthy gentleman farmers who owned large tracts of marshland that were divided into large tenant farms. Advocates of marsh improvement in the first few decades of the nineteenth century were more than twice as wealthy as opponents, who tended to be middling and poor landowners and tenants. Grettler, "Landscape of Reform," 174, 155–197.

52. John Rutherford, "Notes on the State of New Jersey," written in 1776, reprinted in *Proceedings of the New Jersey Historical Society* 1 (1867), 78–89. Quoted in Wacker and Clemens, *Land Use in Early New Jersey,* 121.

53. Sebold, *From Marsh to Farm,* 57–59, Grettler, "Landscape of Reform," 160.

54. Sebold, *Marsh to Farm,* 57–59. In the mid-eighteenth century, notices of intent to petition the New Jersey General Assembly for permission to reclaim marshland frequently appeared in the *Pennsylvania Gazette.* These notices typically employed language like that in the following example: "May 4, 1769. NOTICE is hereby given to whom it may concern, that the owners of a piece or parcel of wild marsh, bounding on Delaware river, in the county of Salem, and province of New Jersey, lying and being between the lands of John Mecum and Allen Congleton, do intend to petition the house of General Assembly of the said province, for leave to bring in a bill at the next sessions of General Assembly, for the banking, laying of sluices, and other things needful to be done, towards draining the aforesaid piece of wild marsh." Accessible Archives, *Pennsylvania Gazette,* CD-ROM, May 4, 1769.

55. The story of marsh reclamation in southwestern New Jersey is captured by numerous minute books kept by some of the region's marsh drainage companies and now resting in the collections of the Salem County Historical Society. The minutes of the Mannington Meadows Company are among the most chronologically comprehensive, extending from 1756 to 1828 (MS #70, SCHS). Other early marsh and meadow company records include those kept by the Lower Alloways Creek Bank Company, 1801–1830 (MS #72, SCHS), Wyatt's New Drain Meadow Company, Mannington, 1820 (MS #74, SCHS), and the Fishing Island Bank and Meadow Company, 1829 (MS #11, SCHS).

56. Grettler, *Landscape of Reform*, 174.

57. Worster, *Rivers of Empire*, 7, 20. Karl Wittfogel originated the concept of the hydraulic society. Wittfogel maintained that such societies emerged when control over water resources by a more powerful elite increased. Wittfogel argued that "where the scale of water control escalated in the ancient desert world, . . . where larger and larger dams and more and more elaborate canal networks were built, political power came to rest in the hands of an elite, typically a ruling class of bureaucrats . . . in their most extreme forms they became despotic regimes in which one or a few supreme individuals wielded absolute control over the common people as they did over the rivers that coursed through their territory." Worster, *Rivers of Empire*, 22–23.

58. Worster, *Rivers of Empire*, 31–32, 37, 48.

59. Mart A. Stewart, *"What Nature Suffers to Groe": Life, Labor, and Landscape on the Georgia Coast, 1680–1920* (Athens: University of Georgia Press, 1996), 147–148, 16, 92. See also Mart A. Stewart, "Rice, Water, and Power: Domination and Resistance in the Low Country, 1790–1900," *Environmental History Review* 15 (Fall 1991): 47–64. For more on hydraulic societies, see Denis Cosgrove, "An Elemental Division: Water Control and Engineered Landscape," in Denis Cosgrove and Geoff Petts, eds., *Water, Engineering and Landscape: Water Control and Landscape Transformation in the Modern Period* (London: Belhaven Press, 1990), 1–11; and Robin Butlin, "Drainage and Land Use in the Fenlands and Fen-edge of Northeast Cambridgeshire in the Seventeenth and Eighteenth Centuries," in Cosgrove and Petts, *Water, Engineering and Landscape*, 54–76.

60. Stewart, *"What Nature Suffers to Groe,"* 104, 148.

61. As cited in Stewart, *"What Nature Suffers to Groe,"* 149. Roswell King, Jr., to Thomas Butler, May 27, 1827. Butler Family Papers, Historical Society of Pennsylvania.

62. Chiarappa, "'The First and Best Sort,'" 92.

63. 1798 federal direct tax, Lower Alloways Creek and Mannington townships, Salem County, New Jersey. In Lower Alloways Creek, the assessors failed to list construction materials and dimensions for a significant percentage of the houses they enumerated. From the information that is provided, the proportion of brick houses compared to all other types of houses resembles that of Mannington and Pittsgrove.

64. The direct tax record for Mannington is incomplete because only the "A" list—that portion of the tax that enumerated houses and lots valued at more than $100—survives. Nevertheless, adjacent Lower Alloways Creek, for which both the "A" and the "B" lists (the latter enumerates houses valued at less than $100 and lots of more than one acre) survive, provides a valid and nearby comparison.

65. Julie Riesenweber studied a group of Salem County's room-by-room probate in-

ventories from 1700 to 1774 and determined that the region's built environment consisted of a fairly limited range of house types. Riesenweber found that the most common house forms consisted of open plans—one-room, hall-parlor, and three- and four-room plans—rather than more modern closed Georgian-plan houses that featured entry into a segregated passage. Julie Riesenweber, "Order in Domestic Space: House Plans and Room Use in the Vernacular Dwellings of Salem County, New Jersey, 1700–1774" (master's thesis, University of Delaware, 1984).

66. Bernard L. Herman, *Architecture and Rural Life in Central Delaware, 1700–1900* (Knoxville: University of Tennessee Press, 1987), 15–17. Lanier and Herman, *Everyday Architecture,* 12–16.

67. Sickler, *Old Houses of Salem County,* 45–48. Kiger House/Jesuit Mission clippings file, SCHS, Historic American Building Survey NJ-445.

68. Although the house now stands a full two stories high and appears at first glance to have always been a two-story building, it is listed on the 1798 tax as a one-and-one-half-story dwelling. A clear shift in the brick bonding and a diagonal disturbance and depression on both the interior and exterior surfaces of the west gable suggest that some sort of major alteration may have occurred there, but the evidence is inconclusive. Like the William Hall house and so many of the other eighteenth-century brick houses in southwestern New Jersey, the Kiger House stood on a major water transportation route; thus, its most public elevation may originally have been the one facing Salem Creek rather than that facing the road. The visible brickwork would seem to contradict this theory, as the present rear elevation lacks a water table and is laid in a much less formal mixture of common and English bond. But upon further inspection, other evidence suggests the possibility that the entire rear wall of the house may have been rebuilt. At only 10 inches, or one brick, thick, the rear (north) wall is much thinner than the south (front) wall, which measures a full 1 foot 3 inches thick—the cumulative width of a brick stretcher, a header, the mortar joint between, and the interior coating of plaster and lath, and a common width for exterior brick walls.

69. In 1797, William Nicholson's assessed value was £150, well above the mean of £42 for the total population, of £72 for the landholding population, and even above the mean of £128 for the owners and occupants of brick houses. Similarly, his house was more highly valued than the house of the average landowner. Mannington 1797 local tax and Mannington 1798 direct tax.

70. Although the lone surviving portion of the 1798 tax record for Mannington—the "A" list—implies that the Nicholson house consisted only of the original 20-by-25-foot chambered hall plan and that the kitchen addition had not yet been built, architectural evidence confirms the kitchen's early- to mid-eighteenth-century construction date. This obvious discrepancy suggests either that the tax assessor may not have included the measurements for attached kitchen wings because he considered them to be unrelated to the main dwelling or that he may have itemized them as separate buildings on the missing "B" list.

71. Hall-parlor plan dwellings accounted for more than half of Julie Riesenweber's sample of Salem County probate inventories between 1700 and 1774. Riesenweber, "Order in Domestic Space."

72. Riesenweber, "Order in Domestic Space."

73. The gambrel roof may have carried associations of status in southwestern New Jersey. Kevin M. Sweeney has argued that such roofs on eighteenth-century Connecticut River Valley dwellings were clearly linked with status and public authority. Gambrel roofs were associated not only with public buildings there but also with the dwellings of justices of the peace, which constituted semipublic buildings because the justices often held court in their homes. Sweeney found that the elite "River Gods" he studied occupied a high proportion of all gambrel-roofed houses in western Massachusetts. Sweeney, "Mansion People."

74. The most obvious change was a new roof and a shift in the upper pitch of the gambrel roof, which necessitated adding several more courses of bricks near the roof peak.

75. Based on this evidence, the existing one-and-one-half-story kitchen wing may have replaced an earlier, lower, one-story kitchen or it may have been the first and only kitchen addition. The evidence from the 1798 tax lists is inconclusive, as it appears that the assessor not only estimated (often incorrectly) the dimensions of most houses but also may have considered attached kitchen wings as separate buildings and enumerated them as such.

76. The Stewart and Naudain houses in Port Penn, Delaware, exhibit similar shifts in brickwork. Bernard L. Herman, personal communication, January 15, 1998.

77. Shourds, *History and Genealogy of Fenwick's Colony*, 93–94.

78. Arthur F. Sewall, *National Register of Historic Places: The Richard Brick House*, New Jersey Historic Sites Inventory 1889.5 (Mannington Township, Salem County, New Jersey), October 1975. Shourds, *History and Genealogy of Fenwick's Colony*, 42–43.

79. Bernard L. Herman, personal communication, February 1998.

80. In his study of surviving pattern brickwork houses in New Jersey, Michael Chiarappa found that double-cell pattern brickwork houses were built only in Gloucester and Burlington Counties, which were closest to the then-major urban environments of Philadelphia and Burlington. Chiarappa, "'The First and Best Sort,'" 145–146.

81. Chiarappa, "'The First and Best Sort,'" 145–146.

82. Cary Carson, Norman F. Barka, William M. Kelso, Garry Wheeler Stone, and Dell Upton, "Impermanent Architecture in the Southern American Colonies," in Robert Blair St. George, ed., *Material Life in America, 1600–1860* (Boston: Northeastern University Press, 1988), 119–121.

83. Herman, *Architecture and Rural Life*, 23–27. About one-quarter of Julie Riesenweber's sample of Salem County probate inventories consisted of three-room plans. These included three types of three-room houses: "Penn plan" houses, those with an entry that was partitioned off of one room and usually smaller, and those that were the result of the addition of one or two rooms to a preexisting hall-parlor or hall-plan house. Riesenweber, "Order in Domestic Space." Chiarappa, "'The First and Best Sort,'" 157–169.

84. Julie Riesenweber's sample of Salem County probate inventories uncovered few Georgian plan houses in use in the eighteenth century. Riesenweber, "Order in Domestic Space." Chiarappa, "'The First and Best Sort,'" 170–199.

85. The house was nominated to the National Register of Historic Places in 1992.

86. Two houses in Gloucester County bear a close resemblance to the Middleton house. The Joseph Thackara House (1754) in Newton Township features a similar spatial arrangement and also contains one small unheated room adjacent to the stair passage. The James Hillman House (1756) in Gloucester Township features a rear stair passage and four rather than three heated rooms. Chiarappa, "'The First and Best Sort,'" 190, 192.

87. Shourds, *History and Genealogy of Fenwick's Colony*, 255.

88. Gowans, "Mansions of Alloways Creek," 381.

89. Shourds, *History and Genealogy of Fenwick's Colony*, 19–21, 35–35, 86, 165, 366, 368, 474.

90. Julia A. King, "'The Transient Nature of All Things Sublunary': Romanticism, History, and Ruins in Nineteenth-Century Southern Maryland," in Rebecca Yamin and Karen Bescherer Metheny, eds., *Landscape Archaeology: Reading and Interpreting the American Historical Landscape* (Knoxville: University of Tennessee Press, 1996), 249–272. See also Julia A. King, "Rural Landscape in the Mid-Nineteenth-Century Chesapeake," in Paul A. Shackel and Barbara J. Little, eds., *Historical Archaeology of the Chesapeake*, (Washington, D.C.: Smithsonian Institution Press, 1994), 283–299.

91. Anne Yentsch, "Legends, Houses, Families, and Myths: Relationships between Material Culture and American Ideology," in Mary C. Beaudry, ed., *Documentary Archaeology in the New World* (Cambridge: Cambridge University Press, 1988), 11.

92. The peculiarity of invented traditions is that they are responses to novel situations that refer to old situations but evince a continuity with the actual past that is largely factitious. Eric Hobsbawm and Terence Ranger, eds., *The Invention of Tradition* (Cambridge: Cambridge University Press, 1983), 1–2. See also Benedict Anderson, *Imagined Communities: Reflections on the Origin and Spread of Nationalism*, rev. ed. (London: Verso, 1991), 173–74, 195.

93. Linda Nochlin, *Women, Art, and Power* (New York: Harper & Row, 1988), 2.

## ⇥ CHAPTER 5 ⇤

1. William Priest, *Travels in the United States of America; Commencing in the year 1793, and ending in 1797* (London: J. Johnson, 1802), 27–30.

2. *William Morgan's Diary and Autobiography*, 240. DSA. Harold B. Hancock, ed., "William Morgan's Autobiography and Diary: Life in Sussex County, 1780–1857" *Delaware History* 19, no. 1 (Spring–Summer 1980): 39–52.

3. Robert G. Johnson, Esq., Salem County, New Jersey, Account Book, 1786 to 1850, 103. SCHS.

4. William Cobbett, *A Years Residence in the United States of America, in three parts* (1818–1819; reprint, New York: Augustus M. Kelley, Publishers, 1969), 64–65.

5. Bernhard, Duke of Saxe-Weimar Eisenach, *Travels Through North America during the Years 1825 and 1826* (Philadelphia: Carey, Lea, & Carey, 1828), 148–149.

6. "An Account of the Progress of Population, Agriculture, Manners, and Govern-

ment in Pennsylvania, in a letter from a citizen of Pennsylvania, to his friend in England," *Columbian Magazine, or Monthly Miscellany* 1, no. 3 (November 1786): 117–122.

7. Michael O'Brien, "On Observing the Quicksand," *American Historical Review* 104, no. 4 (October 1999): 1202–1207. Celia Applegate, "A Europe of Regions: Reflections on the Historiography of Sub-National Places in Modern Times," *American Historical Review* 104, no. 4 (October 1999), 1157–1182. Edward L. Ayers and Peter S. Onuf, "Introduction," in Edward L. Ayers, Patricia Nelson Limerick, Stephen Nissenbaum, and Peter S. Onuf, *All Over the Map: Rethinking American Regions* (Baltimore: Johns Hopkins University Press, 1996), 1–10. Stephen Nissenbaum, "New England as Region and Nation," in Ayers et al., *All Over the Map*, 38–61.

8. For a discussion of the notion of American exceptionalism and its contribution to the formation of an American identity, see Jack P. Greene, *The Intellectual Construction of America: Exceptionalism and Identity from 1492 to 1800* (Chapel Hill: University of North Carolina Press, 1993), 198–199, 200–209.

9. Michael Kammen, *Mystic Chords of Memory: The Transformation of Tradition in American Culture* (New York: Alfred A. Knopf, 1991), 88–89.

10. Focusing on symbolic expressions of national ideology ranging from heroes and place names to holidays and public sculpture, Zelinsky suggested that, because the individuals and communities transplanted into early eastern North America basically replicated Old World models, there was initially little overt evidence of the imprint of a central authority on the landscape. Wilbur Zelinsky, "The Imprint of Central Authority," in Michael P. Conzen, ed., *The Making of the American Landscape* (New York: Routledge, 1990, 1994), 311–334, quote on 311. Wilbur Zelinsky, *Nation into State: The Shifting Symbolic Foundations of American Nationalism* (Chapel Hill: University of North Carolina Press, 1988).

11. In her work on regional identity in East Asia, Kären Wigen has argued that regional identity is unquestionably fluid and constitutes more of a process than a product. A crucial question to ask, she suggests, is *how* and *where* such regional identities are actually produced. While many aspects of regional and national identity may be imposed from without rather than emerging from within, in East Asia, households and families constitute one critical locus for the transmission and transformation of regional identity, through language acquisition, marriage customs, and domestic labor practices. In the study of American regional identity formation, it is also worth considering where and how such identities get formed and renegotiated. Kären Wigen, "Culture, Power, and Place: The New Landscapes of East Asian Regionalism," *American Historical Review* 104, no. 4 (October 1999), 1183–1201. Edward Ayers and Peter Onuf have suggested that local distinctions had always existed in early America, but the critical question for us to be asking is not *what* those distinctions actually were, but *why* they came to matter in the construction of collective identities. Ayers and Onuf, "Introduction" in *All over the Map*, 8.

12. "Manuscript Diary of Richard Wood," *South Jersey Magazine*, Fall 1974–Fall 1984: October 3, 1810, 24.

13. In the latter part of the eighteenth century, Philip Fithian remarked that an acquaintance from Greenwich, New Jersey, "had their vessel overset, laden with 700

bushel of wheat near Reedy Island" just off the Delaware coast. F. Alan Palmer, *The Beloved Cohansie of Philip Vickers Fithian* (Greenwich, N.J.: Cumberland County Historical Society, 1990), 79.

14. "Manuscript Diary of Richard Wood."

15. François Alexandre Frédéric, duc de la Rochefoucauld-Liancourt, *Travels through the United States of North America, the country of the Iroquois, and upper Canada, in the years 1795, 1796, 1797; with an authentic account of Lower Canada*, 2 vols. (London: R. Phillips, 1799), 1:42.

16. "Manuscript Diary of Richard Wood."

17. Francis Baily, *Journal of a Tour in Unsettled Parts of North America in 1796 & 1797*. (London: Baily Brothers, 1856), 107–108. Palmer, *The Beloved Cohansie of Philip Vickers Fithian*, 187, 198.

18. In North West Fork Hundred, Delaware, the Maston House (Figure 3.2), which Henry Chandlee Forman called a "Maryland house in Delaware," features chevron-patterned brickwork on its gables. In Maryland, see, for example, St. Paul's Church and Vestry House and the Violet Farm, Fairlee vicinity, Kent County, Maryland; Providence Plantation, Quaker Neck, Kent County, Maryland; and Cloverfields, Wye Mills vicinity, Queen Anne's County, Maryland. Michael Bourne, Orlando Ridout V, Paul Touart, and Donna Ware, *Architecture and Change in the Chesapeake: A Field Tour on the Eastern and Western Shores* (Crownsville, Md.: Vernacular Architecture Forum and the Maryland Historical Trust Press, 1998), 69–73, 76–77, 83–84.

19. The primary core region for the Pennsylvania barn extends well north and west of Delaware and continues into Virginia. Robert F. Ensminger, *The Pennsylvania Barn: Its Origins, Evolution, and Distribution in North America* (Baltimore: Johns Hopkins University Press, 1992), 51–52. The S. H. Rothwell Farm Barn and the Wheatland Farm Barn, both in St. Georges Hundred, in southern New Castle County, Delaware, illustrate this process of architectural negotiation in agricultural buildings. Gabrielle M. Lanier et al., *Threatened Resources Documented in Delaware, 1991–92* (Newark, Del.: Center for Historic Architecture and Engineering, 1992), 21–27. Bernard L. Herman, *Architecture and Rural Life in Central Delaware, 1700–1900* (Knoxville: University of Tennessee Press, 1987), 214–215, 222. Gabrielle M. Lanier and Bernard L. Herman, *Everyday Architecture of the Mid-Atlantic: Looking at Buildings and Landscapes* (Baltimore: Johns Hopkins University Press, 1997), 199–201.

20. Cynthia Gayle Falk, "Evidence of Ethnicity and Status in the Architectural Landscape of Eighteenth-Century Coventry Township, Chester County, Pennsylvania." (master's thesis, University of Delaware, 1996). Nancy Van Dolsen, *Cumberland County: An Architectural Survey* (Carlisle, Pa.: Cumberland County Historical Society, 1990), 13–19.

21. Ferdinand-M. Bayard, *Travels of a Frenchman in Maryland and Virginia with a description of Philadelphia and Baltimore in 1791 or Travels in the Interior of the United States, to Bath, Winchester, in the Valley of the Shenandoah, etc., etc., during the Summer of 1791* (Ann Arbor, Mich.: Edwards Brothers, 1950), 98. Henry Bradshaw Fearon, *Sketches of America* (London: Strahan & Spottiswoode, 1819), 183–184.

22. Rayner Wickersham Kelsey, ed., *Cazenove Journal 1794: A Record of the Journal of*

*Theophile Cazenove through New Jersey and Pennsylvania* (Haverford, Pa.: Pennsylvania History Press, 1922), 81.

23. Lanier and Herman, *Everyday Architecture,* 83–84, 86, 90–91, 189.

24. Historic American Buildings Survey, Caesaria River House (Wurster House), Greenwich Township, Cumberland County, New Jersey. HABS NJ-986, Gabrielle M. Lanier, delineator. For Rhode Island linkages, see Myron O. Stachiw et al., *The Early Architecture and Landscapes of the Narragansett Basin,* Vol. 1: *Newport* (Newport, R.I.: Vernacular Architecture Forum, 2001).

25. Gail Hunton and James C. McCabe, *Monmouth County Historic Sites Inventory: Summary Report* (Lincroft, N.J.: Office of New Jersey Heritage, Monmouth County Park System, and the Monmouth County Historical Association, 1980–1994, 1990), 60–67.

26. Falk, "Evidence of Ethnicity and Status." See also Cynthia G. Falk, "Symbols of Assimilation or Status? The Meanings of Eighteenth-Century Houses in Coventry Township, Chester County, Pennsylvania," *Winterthur Portfolio* 33, no. 2/3 (1998): 107–134.

# PRIMARY SOURCES

## → ARCHITECTURE ←

What follows is a partial listing of buildings and landscapes examined and reviewed during the course of this study. For a fuller listing of available architectural resources, both extant as well as surveyed but now demolished, consult the Cultural Resource Survey files in the state historic preservation office for each area.

### Lancaster County, Pennsylvania

Christian Becker House (Warwick), Christian Eby House (Elizabeth), Christian Eby House (Warwick), Philip and Barbara Friedrich House (Warwick), Johannes Hess House (Lititz), Peter Kreiter House (Lititz), John and Elizabeth Pfautz House (Warwick), Reist House (Warwick), Rudy House (Warwick), Warden's House (Lititz), Warner House (Lititz), Weaver House (West Lampeter).

### Salem County, New Jersey

James Barrett House (Mannington), Richard Brick House (Mannington), John Maddox Denn House (Lower Alloways Creek), Jacob Fox House (Mannington), William Hall House (Mannington), William Hancock House (Lower Alloways Creek), Hedgefield (Mannington), Kiger House or Jesuit Mission (Mannington), Hippolyte LeFevre House (Mannington), Lower Alloways Creek Friends Meeting House (Lower Alloways Creek), Hugh Middleton House (Mannington), William Nicholson House (Mannington), William Oakford House (Alloway), John Pledger Jr. House (Mannington), Poplar Tree Farm (Mannington), Richman House (Mannington), William Smith House (Mannington).

### Sussex County, Delaware

Cannon Hall (Cannon's Ferry), Maston House (North West Fork Hundred), Sudler House (North West Fork Hundred), White-Brown House (North West Fork Hundred).

## →¦ MANUSCRIPTS AND IMAGES ¦←

*See list of abbreviations in Notes.*

Brooke, Robert. Survey Notes, 2 vols. MS #122: AM 26711. HSP.

Delaware Orphans Court Records: New Castle, Kent, and Sussex Counties. DSA.

Fishing Island Bank and Meadow Company Minute Book, 1829. MS #11, SCHS.

Johnson, Robert G., Esq. Account Book, 1786–1850. MS #5, SCHS.

Lower Alloways Creek Bank Company Minute Book, 1801–1830. MS #72, SCHS.

Mannington Meadows Company Minute Book, 1756–1828. MS #70, SCHS.

Mannington Township, Salem County, New Jersey, Local Tax Lists. SCHS.

Morgan, William. Diary and autobiography. DSA.

North West Fork Hundred, Sussex County, Delaware, Local Tax Assessments for 1801, 1803, and 1816. DSA.

Probate Inventories, Chester County, Pennsylvania. Chester County Archives (CCA).

Probate Inventories, Lancaster County, Pennsylvania. LCHS.

Probate Inventories, Salem County, New Jersey. SCCH.

Probate Inventories, Sussex County, Delaware. DSA.

Sadsbury Township, Chester County, Pennsylvania, Local Triennial Tax Lists, 1796–1820. CCA.

Shankland Surveys. Sussex County Recorder of Deeds, Surveys A (1776) and Surveys B (1776). DSA.

Simkins, J. H. Paintings. SCHS.

Sussex County Coroner's Reports. DSA.

Sussex County Wills. DSA.

United States Census, Delaware Manuscript Returns, Sussex County, 1800, 1810, 1820, 1830. DSA.

United States Coastal Survey Maps, 1841 and 1882. National Archives.

United States Direct Tax, 1798. Chester County, Pennsylvania: Sadsbury and Tredyffrin Townships; Lancaster County, Pennsylvania: Conestoga, Hempfield, and Warwick Townships; Philadelphia: Northern Liberties; Salem County, New Jersey: Lower Alloways Creek, Mannington, and Pittsgrove Townships.

United States Direct Tax, 1815. Lancaster County, Pennsylvania: Conestoga, Hempfield, and Warwick Townships.

Whitelaw, James. *Journal, 1773–93.* Manuscript B WS8, Vermont Historical Society.

Wilson, Samuel. Account Books, 1780–1822. 3 vols. CCHS.

Wyatt's New Drain Meadow Company Minute Book, 1820. MS #74, SCHS.

Yorke, Thomas. Photograph album. SCHS.

## →¦ PUBLISHED PRIMARY SOURCES ¦←

Accessible Archives, *Niles Register: Cumulative Index, 1811–1849,* CD-ROM. Malvern, Pa.: Accessible Archives, 1995.

Accessible Archives, *Pennsylvania Gazette* CD-ROM, Folios 1, 2, and 3. Provo, Utah: Folio Corp., 1991.

*American Museum or Repository of Ancient and Modern Fugitive Pieces, & c. Prose and Poetical,* For April 1788. Philadelphia: Carey, Stewart, & Co. 1790.

*American Universal Magazine* 3, no. 1 (Monday, July 10, 1797).

"An Account of the Progress of Population, Agriculture, Manners, and Government in Pennsylvania, in a letter from a citizen of Pennsylvania, to his friend in England," *The Columbian Magazine, or Monthly Miscellany* 1, no. 3 (November 1786): 117–122.

Anburey, Thomas. *Travels through the interior parts of America: In a series of letters. By an officer.* London: Printed for William Lane, 1789.

Baily, Francis, F.R.S. *Journal of a Tour in Unsettled Parts of North America in 1796 & 1797.* London: Baily Brothers, 1856.

Bayard, Ferdinand-M. *Travels of a Frenchman in Maryland and Virginia with a description of Philadelphia and Baltimore in 1791 or Travels in the Interior of the United States, to Bath, Winchester, in the Valley of the Shenandoah, etc., etc., during the Summer of 1791.* Ann Arbor, Michigan: Edwards Brothers, 1950.

Benson, Adolph B., ed. and trans. *Peter Kalm's Travels in North America: The English Version of 1770.* New York: Wilson-Erickson, 1937.

Bordley, John Beale. *Essays and Notes on Husbandry and Rural Affairs.* Philadelphia: Printed by Budd and Bartram, for Thomas Dobson at the Stone House, no. 41, South Second Street, 1799.

Brissot de Warville, Jacques Pierre. *New Travels in the United States of America. In Two Volumes.* London: J. S. Jordan, 1794.

Chauncey, Isaac, et al. "New Jersey Salt Marsh Company Charter." *American Farmer* 2, no. 20 (August 11, 1820), 153–156.

Chotanter, A. "The Draining of Marshes." *American Farmer* 2 (October 27, 1820), 243–245.

Cobbett, William. *A Year's Residence in America.* Boston: Small, Maynard & Co., 1922.

Cooper, Thomas. *Some Information Respecting America.* London: printed for J. Johnson, 1794.

Currie, William. "Face of the Country: Soil, Climate, and Productions of Delaware." *Delaware Register and Farmers' Magazine* 1, no. 2 (1838), 110–112.

Currie, William. *An Historical Account of the Climate and Diseases of the United States of America.* Philadelphia: Dobson, 1792.

Darby, James C. "On the Embanking and Preparation of Marsh Land, for the Cultivation of Rice." *Southern Agriculturist, Horticulturist, and Register of Rural Affairs* 2 (January 1829): 23–28.

*Delaware Gazette and American Watchman.* Wilmington: Samuel Harker, 1828–1837.

Douglass, Frederick. *My Bondage and My Freedom.* Urbana: University of Illinois Press, 1987.

"Draining." *American Farmer* 8, no. 21 (August 11, 1826), 162–163.

Farish, Hunter D., ed. *Journal and Letters of Philip Vickers Fithian, 1773–1774.* Williamsburg, Va.: Colonial Williamsburg, 1957.

Fearon, Henry Bradshaw. *Sketches of America.* London: Strahan and Spottiswoode, 1819.

Fithian, Philip Vickers. "Journal and Letters, 1766–1767." Special Collections, Firestone Library, Princeton University.

"Inland Swamps." *American Farmer* 5 (August 1, 1823), 147.

Johnson, Robert G., Esq., Account Book, 1786–1850. MS #5, SCHS.

Johnson, Robert G. *An Historical Account of the First Settlement of Salem in West Jersey by John Fenwick, Esq., Chief Proprietor of the Same.* Philadelphia: O. Rogers, 1839.

Johnson, Robert Gibbon. "On Reclaiming Marsh Land." *American Farmer* 8, no. 24 (September 1, 1826), 185–187; *American Farmer* 8, no. 25 (September 8, 1826): 201–202; *American Farmer* 8, no. 26 (September 15, 1826): 185–187.

*Journals of Congress, VII.* Philadelphia: R. Aitken, 1777–1788.

Kelsey, Rayner Wickersham, ed. *Cazenove Journal, 1794: A Record of the Journey of Theophile Cazenove through New Jersey and Pennsylvania.* Haverford: Pennsylvania History Press, 1922.

La Rochefoucauld-Liancourt, François Axexandre Frédéric, duc de. *Travels through the United States of North America, the country of the Iroquois, and upper Canada, in the years 1795, 1796, 1797; with an authentic account of Lower Canada.* 2 vols. London: R. Phillips, 1799.

Lewis, Enoch, ed. *The African Observer: A Monthly Journal Containing Essays and Documents Illustrative of the General Character and Moral and Political Effects, of Negro Slavery.* Westport, Conn.: Negro Universities Press, 1970.

Minor, P. "No. 3—On Draining." *American Farmer* 1, no. 29 (December 24, 1819), 308–9.

*Monthly Magazine, and American Review* for 1799, vol 1. New-York: T. & J. Swords, 1800.

Moraley, William. *The Infortunate: or the Voyage and Adventures of William Moraley.* Newcastle[, U.K.?]: J. White, 1743.

Morse, Jedediah. *The American Universal Geography.* Boston: Buckingham for Thomas & Andrews, 1805.

*Narrative and Confessions of Lucretia P. Cannon, Who was Tried, Convicted, and Sentenced to be Hung at Georgetown, Delaware, with Two of Her Accomplices. Containing an Account of Some of the Most Horrible and Shocking Murders and Daring Robberies Ever Committed by One of the Female Sex.* New York: New York Publishers, 1841.

*Niles Weekly Register.* Baltimore: Hezekiah Niles, 1814–1837.

"On Draining." *American Farmer* 7, no. 9 (May 20, 1825), 67–68.

"On Draining Grass or Meadow Land." *American Farmer* 7, no. 9 (May 20, 1825), 67.

"On Gentleman Farming." *American Farmer* 8, no. 44 (January 19, 1827), 345–346; and *American Farmer* 8, no. 45 (January 26, 1827), 354–356.

"On Reclaiming Marsh Land." *American Farmer* 2 (July 21, 1820), 131–132.

Palmer, F. Alan, ed. *The Beloved Cohansie of Philip Vickers Fithian (1747–1776).* Greenwich, N.J.: Cumberland County Historical Society, 1990.

Priest, William. *Travels in the United States of America, Commencing in the Year 1793, and ending in 1797.* London: J. Johnson, 1802.

Read, George, ed. *Laws of the State of Delaware.* New Castle: printed by Samuel and John Adams, 1797–1816.

Roberts, Kenneth, and Roberts, Anna M., eds. and trans. *Moreau de St. Mery's American Journey [1793–1798].* Garden City, N.Y.: Doubleday, 1947.

Rush, Benjamin. "An Account of the Manners of the German Inhabitants of Pennsylvania." *Columbian Magazine*, January 1789, 22–30.

Rush, Benjamin. *Essays Literary, Moral, and Philosophical.* Edited and introduced by Michael Meranze. Schenectady, N.Y.: Union College Press, 1988.

Rusticus, Philo. "Inland Swamps." *American Farmer* 5, no. 19 (August 1, 1823), 147.

Rutherford, John. "Notes on the State of New Jersey" (1776). In *Proceedings of the New Jersey Historical Society* 1 (1867), 78–89.

Schoepf, Johann. *Travels in the Confederation 1783–1784.* Translated and edited by Alfred J. Morrison. Philadelphia: William J. Campbell, 1911.

Skinner, J. S., Esq. "Reclaimed Marsh." *American Farmer* 9, no. 33 (December 7, 1827): 300–301.

Swartout, Samuel. "On Reclaiming Salt Marshes." *American Farmer* 1, no. 25 (November 26, 1819), 277–278.

Thompson, William, Esq. "Method of Reclaiming Swamp Land." *American Farmer* 9, no. 14 (June 22, 1827), 106–107.

Torrey, Jesse. *American Slave Trade.* London: J. M. Cobbett, 1822.

Townsend, George Alfred. *The Entailed Hat or Patty Cannon's Times.* New York: Harper & Brothers, Franklin Square, 1884.

Webster, Noah. "On the Regularity of the City of Philadelphia." In *A Collection of Essays and Fugitive Writings on Moral, Historical, Political and Literary Subjects,* 217–221. Boston: I. Thomas and E. T. Andrews, 1790.

*Weekly Magazine.* Philadelphia: James Watters & Co., 1798–1799.

Weld, Isaac. *Travels Through the United States of North America and the Provinces of Upper and Lower Canada, during the Years 1795, 1796, and 1797.* London: printed for J. Stockdale, 1800.

*West Jersey Gazette and Salem and Gloucester Advertiser.* Salem: Isaac A. Kollock, 1817–1819. SCHS.

Wood, Richard. Manuscript Diary, 1801–1822, Cumberland County, New Jersey. *South Jersey Magazine,* Fall 1974–Fall 1984.

Wright, Charles. "Near Seaford, Sussex County, Delaware, December 3, 1851." Patent Office Report, 1852, vol. 2: H. Document #102, p. 263.

*Page numbers in italics refer to illustrations.*